Georg Hulla

Low energy ion induced desorption on technical surfaces

Georg Hulla

Low energy ion induced desorption on technical surfaces

At room temperature

Südwestdeutscher Verlag für Hochschulschriften

Impressum/Imprint (nur für Deutschland/ only for Germany)
Bibliografische Information der Deutschen Nationalbibliothek: Die Deutsche Nationalbibliothek verzeichnet diese Publikation in der Deutschen Nationalbibliografie; detaillierte bibliografische Daten sind im Internet über http://dnb.d-nb.de abrufbar.
Alle in diesem Buch genannten Marken und Produktnamen unterliegen warenzeichen-, marken- oder patentrechtlichem Schutz bzw. sind Warenzeichen oder eingetragene Warenzeichen der jeweiligen Inhaber. Die Wiedergabe von Marken, Produktnamen, Gebrauchsnamen, Handelsnamen, Warenbezeichnungen u.s.w. in diesem Werk berechtigt auch ohne besondere Kennzeichnung nicht zu der Annahme, dass solche Namen im Sinne der Warenzeichen- und Markenschutzgesetzgebung als frei zu betrachten wären und daher von jedermann benutzt werden dürften.

Verlag: Südwestdeutscher Verlag für Hochschulschriften Aktiengesellschaft & Co. KG
Dudweiler Landstr. 99, 66123 Saarbrücken, Deutschland
Telefon +49 681 37 20 271-1, Telefax +49 681 37 20 271-0, Email: info@svh-verlag.de
Zugl.: Vienna, University of Technology, Dissertation, 2009

Herstellung in Deutschland:
Schaltungsdienst Lange o.H.G., Berlin
Books on Demand GmbH, Norderstedt
Reha GmbH, Saarbrücken
Amazon Distribution GmbH, Leipzig
ISBN: 978-3-8381-0760-8

Imprint (only for USA, GB)
Bibliographic information published by the Deutsche Nationalbibliothek: The Deutsche Nationalbibliothek lists this publication in the Deutsche Nationalbibliografie; detailed bibliographic data are available in the Internet at http://dnb.d-nb.de.
Any brand names and product names mentioned in this book are subject to trademark, brand or patent protection and are trademarks or registered trademarks of their respective holders. The use of brand names, product names, common names, trade names, product descriptions etc. even without a particular marking in this works is in no way to be construed to mean that such names may be regarded as unrestricted in respect of trademark and brand protection legislation and could thus be used by anyone.

Publisher:
Südwestdeutscher Verlag für Hochschulschriften Aktiengesellschaft & Co. KG
Dudweiler Landstr. 99, 66123 Saarbrücken, Germany
Phone +49 681 37 20 271-1, Fax +49 681 37 20 271-0, Email: info@svh-verlag.de

Copyright © 2009 by the author and Südwestdeutscher Verlag für Hochschulschriften Aktiengesellschaft & Co. KG and licensors
All rights reserved. Saarbrücken 2009

Printed in the U.S.A.
Printed in the U.K. by (see last page)
ISBN: 978-3-8381-0760-8

Contents

1 Introduction and Overview   1

2 CERN   3
- 2.1 Organization 3
- 2.2 Scientific achievements 3
- 2.3 Accelerator complex 4

3 The LHC Project   7
- 3.1 An introduction of the LHC beam vacuum system 10
- 3.2 Beam related dynamic vacuum effects and their impact on LHC 13
  - 3.2.1 Desorption by synchroton radiation 13
  - 3.2.2 Ion-induced pressure instability 14
  - 3.2.3 Beam induced electron multipacting 15

4 Theoretical Framework   19
- 4.1 Ion-induced desorption mechanism 19
- 4.2 Calculation of the desorption yield 20
- 4.3 Calculation of the desorption cross section 22
- 4.4 Minimum measurable desorption yield 23
- 4.5 Energy loss of ions passaging through matter 23
  - 4.5.1 Nuclear energy loss 25
  - 4.5.2 Electronic energy loss 28
- 4.6 Sputter yield 29

5 Experimental Setup   31
- 5.1 Ion generation and beam optics 31
  - 5.1.1 Ion gun and first lens 31
    - 5.1.1.1 Ionization probability 31
    - 5.1.1.2 Spark discharge - Paschen curve 34
    - 5.1.1.3 Assembling and function 35
  - 5.1.2 Dipole magnet 36
    - 5.1.2.1 Dipole geometry 37
    - 5.1.2.2 Focusing effect 38
  - 5.1.3 Second lens 40
  - 5.1.4 Beam monitoring and deflection 40
- 5.2 Differential pumping system 41
- 5.3 System calibration 42
  - 5.3.1 Calibration of the pumping speed 42
  - 5.3.2 RGA calibration 45

		5.3.2.1	Sensitivity factors and evaluation of ion currents	47
	5.4	Modifications and additions to the setup	48
		5.4.1	Design of a new ion gun power supply	48
		5.4.2	Pumping speed reduction in the UHV-chamber	50
		5.4.3	Faraday cup for beam positioning	51
		5.4.4	Ion optic simulations with the SIMION program	52
	5.5	Measurement procedure	53
		5.5.1	Energy and mass dependent measurements	55
		5.5.2	Dose dependent measurements	56

6 Results and Discussion 57

	6.1	Results	57
		6.1.1 Influence of the ion nature on the desorption yield	57
		6.1.2 Energy and mass dependance of the desorption yield	58
		6.1.2.1 Noble gas ions	58
		6.1.2.2 Hydrogen containing ions	62
		6.1.2.3 Other type of ions: Oxygen containing ions and nitrogen ions . . .	65
		6.1.3 Dose dependance of the desorption yield	69
		6.1.3.1 Various ions incident on OFHC-copper	69
		6.1.3.2 Ar^+-ions incident on different target materials	73
		6.1.4 Desorption cross section and total coverage	76
		6.1.4.1 Noble gas ions incident on OFHC-copper	76
		6.1.4.2 Ar^+-ions incident on different target materials	78
	6.2	Discussion	79
		6.2.1 Correlation between sputter- and desorption yields	79
		6.2.2 Different reasons for the desorption of molecules with different masses . . .	83
		6.2.3 Prediction of low energy ion desorption yields	85
		6.2.3.1 Noble gas ions	85
		6.2.3.2 Hydrogen containing ions	86
		6.2.3.3 Other type of ions: Oxygen containing ions and nitrogen ions . . .	88
		6.2.4 Influence of the beam shape on the desorption yield	90
		6.2.5 Other effects during desorption	92
		6.2.6 Estimation of the measurement accuracy	93

7 Conclusion 95

A Appendix 97

	A.1	Copper cleaning procedure	97
	A.2	Bake-out procedure	98
	A.3	Stopping units	98
	A.4	Compilation of ion impact desorption cross-section	99
	A.5	Source electronics	100

Bibliography 105

List of Figures

2.1	Member states of CERN.	4
2.2	The accelerator complex at CERN.	5
3.1	General layout of the Large Hadron Collider.	8
3.2	Cross section of a LHC dipole cryomagnet assembly.	10
3.3	Picture of a prototype beam screen, inserted into a sample beam pipe.	12
3.4	Typical plots of the secondary electron yield of copper as a function of the primary electron energy.	18
4.1	Ion-induced desorption mechanism.	20
4.2	Various limitations to η_{min}.	24
4.3	Stopping cross sections for Ar^+-ions incident on copper.	26
5.1	Schema of the experimental setup.	32
5.2	Ion optic simulation of the experimental system.	32
5.3	Single ionization energy of the elements as a function of Z.	33
5.4	Single-, double- and triple ionization cross section as a function of the electron energy.	34
5.5	Paschen curves for various gases.	35
5.6	Schema of the ion gun.	36
5.7	Potential energy view in the ion gun.	37
5.8	Geometry of the dipole magnet.	38
5.9	Deflection angle of the dipole magnet.	39
5.10	Focusing and de-focusing effect of a dipole magnet.	39
5.11	Ion gun settings for Ne^+-ions.	40
5.12	Schema of the differential pumping system.	42
5.13	Schematic of the UHV-chamber.	43
5.14	Measurement of the k-factor.	44
5.15	Calculated pumping speed in the case of argon injection.	45
5.16	Measured and fitted values of the pumping speed in the case of argon injection.	46
5.17	RGA sensitivity during argon injection.	47
5.18	Comparison of BA-gauge and ion gun.	49
5.19	Electronic characteristics in the ion gun without repeller.	49
5.20	Requirements for the ion gun power supply.	50
5.21	Cross section of the Faraday cup.	52
5.22	Beam control rack.	54
5.23	Dose dependance of the desorption yield during ion bombardment.	55
5.24	Short ion pulse desorption signals.	56

6.1 H_2- and CO desorption yields of three different ion types incident on an OFHC-copper sample. 57
6.2 H_2 desorption yields of noble gas ions incident on copper as function of the ion energy. 59
6.3 H_2 desorption yields of noble gas ions incident on copper as function of the ion mass. 59
6.4 CO desorption yields of noble gas ions incident on copper as function of the ion energy. 60
6.5 CO desorption yields of noble gas ions incident on copper as function of the ion mass. 60
6.6 CO_2 desorption yields of noble gas ions incident on copper as function of the ion energy. 61
6.7 CO_2 desorption yields of noble gas ions incident on copper as function of the ion mass. 61
6.8 H_2 desorption yields of hydrogen containing ions incident on copper as function of the ion energy. 62
6.9 H_2 desorption yields of hydrogen containing ions incident on copper as function of the ion mass. 63
6.10 CO desorption yields of hydrogen containing ions incident on copper as function of the ion energy. 63
6.11 CO desorption yields of hydrogen containing ions incident on copper as function of the ion mass. 64
6.12 CO_2 desorption yields of hydrogen containing ions incident on copper as function of the ion energy. 64
6.13 CO_2 desorption yields of hydrogen containing ions incident on copper as function of the ion mass. 65
6.14 H_2 desorption yields of N_2^+-ions and oxygen containing ions incident on copper as function of the ion energy. 66
6.15 H_2 desorption yields of N_2^+-ions and oxygen containing ions incident on copper as function of the ion mass. 66
6.16 CO desorption yields of N_2^+-ions and oxygen containing ions incident on copper as function of the ion energy. 67
6.17 CO desorption yields of N_2^+-ions and oxygen containing ions incident on copper as function of the ion mass. 67
6.18 CO_2 desorption yields of N_2^+-ions and oxygen containing ions incident on copper as function of the ion energy. 68
6.19 CO_2 desorption yields of N_2^+-ions and oxygen containing ions incident on copper as function of the ion mass. 68
6.20 Dose dependance of 7keV He^+-ions incident on copper. 70
6.21 Dose dependance of 7keV Ne^+-ions incident on copper. 70
6.22 Dose dependance of 7keV Ar^+-ions incident on copper. 71
6.23 Dose dependance of 7keV Kr^+-ions incident on copper. 71
6.24 Dose dependance of 7keV N_2^+-ions incident on copper. 72
6.25 Dose dependance of 7keV CO^+-ions incident on copper. 72
6.26 Dose dependance of 7keV Ar^+-ions incident on stainless steel. 73
6.27 Dose dependance of 7keV Ar^+-ions incident on $1\mu m$ copper coated stainless steel. . 74
6.28 Dose dependance of 7keV Ar^+-ions incident on $1\mu m$ silver coated stainless steel. .. 74
6.29 Dose dependance of 7keV Ar^+-ions incident on $1\mu m$ gold coated stainless steel. 75
6.30 Dose dependance of 7keV Ar^+-ions incident on beam screen copper. 75
6.31 Fit result for the calculation of σ and N_0. 76
6.32 Desorption cross section σ obtained for 3 and 7keV noble gas ions incident on copper. 77
6.33 Initial coverage N_0 obtained for 3 and 7keV noble gas ions incident on copper. 77
6.34 Desorption cross section obtained for 7keV Ar^+-ions incident on different target materials. 78

6.35	Initial coverage obtained for 7keV Ar^+-ions incident on different target materials.	78
6.36	Calculated sputter yields versus measured desorption yields for noble gas ions incident on copper.	79
6.37	Ratio of electronic to nuclear energy loss for H_2^+- and noble gas ions incident on copper.	80
6.38	Ion trajectories of He^+- and Kr^+-ions incident on copper.	81
6.39	Microscopic view of an irradiated and a non-irradiated OFHC-copper sample.	82
6.40	H_2 desorption yields of He^+-ions.	84
6.41	H_2 desorption yields of H_2^+-ions.	84
6.42	H_2 desorption yields as a function of the total energy loss obtained for noble gas ions incident on copper.	85
6.43	CO desorption yields as a function of the total energy loss obtained for noble gas ions incident on copper.	86
6.44	CO_2 desorption yields as a function of the total energy loss obtained for noble gas ions incident on copper.	86
6.45	H_2 desorption yields as a function of the total energy loss obtained for hydrogen containing ions incident on copper.	87
6.46	CO desorption yields as a function of the total energy loss obtained for hydrogen containing ions incident on copper.	87
6.47	CO_2 desorption yields as a function of the total energy loss obtained for hydrogen containing ions incident on copper.	88
6.48	H_2 desorption yields as a function of the total energy loss obtained for N_2^+-ions and oxygen containing ions incident on copper.	88
6.49	CO desorption yields as a function of the total energy loss obtained for N_2^+-ions and oxygen containing ions incident on copper.	89
6.50	CO_2 desorption yields as a function of the total energy loss obtained for N_2^+-ions and oxygen containing ions incident on copper.	89
6.51	Subdivision of the beam area into small slices with an assumed Gauss ion density distribution.	91
6.52	Total desorption yield of a beam with a Gauss ion density distribution as function of the ion dose.	91
A.1	Compilation of ion impact desorption cross section.	99
A.2	Overview of the board components.	101
A.3	Board components of grid and extraction.	102
A.4	Board components of the filament.	103
A.5	Picture of the finalized board.	104

List of Tables

5.1	RGA sensitivities for injected gases and their cracking patterns.	48
5.2	Minimum achievable desorption coefficients calculated for various gases.	51
A.1	Multiplication factors for the stopping units of different ions incident on copper. . . .	98

Chapter 1

Introduction and Overview

The ion-induced desorption (IID) was for a long time not a subject for studies but an old recipe to clean surfaces [1]. A strong interest in the release of gases by ion bombardment came in the 1970's. Simultaneously in the accelerator and the fusion community, this release was identified as a main limitation in the performance of storage rings [2] and tokamaks [3].

Although sputtering was already discovered by W.R. Grove in 1852 [4], in 1974 H.F. Winter and P. Sigmund [5] showed that the classical theory of sputtering [6] can partly explain the desorption of chemisorbed gases by low energy ions. The validity of this approach was confirmed experimentally [7] with some restriction in the case of technical surfaces[1] due to the absence of a well defined binding energy for the desorbed molecules [8]. Furthermore other effects related to the energy deposition of the primary particle such as defect creation, stimulated diffusion and excitation of electrons can influence the release of impurities contained in the solid. Clearly the refinements of models used to study the desorption of pure gases from monocrystals are of little use in the case of technical surfaces.

Surface physics investigations of technical materials were successfully launched at that time [9, 10, 11, 12] in order to limit the detrimental consequences of IID, they contributed to define a set of cleaning treatments [13, 14], mainly based on ion bombardment, which resulted in a significant improvement for the performance of Tokamaks [15] and storage rings (in the ISR the stored currents raised from 20 to 60A [16]).

In case of technical surfaces prepared as accelerator vacuum chambers, e.g. degreasing followed by alkaline etching and rinsing in demineralized water, the following investigations on the IID-yield have been carried out:

- The variation of the desorption yield as a function of the ion energy shows that the desorption yield increases smoothly with the ion energy. The maximum is reached in the keV region and threshold energies close to 1eV have been measured [17, 18, 19].

---

[1]The surface contains a lot of physical and chemical imperfections, which makes predictions of desorption yields and cross sections difficult.

- The variation of the desorption yield with the surface treatment shows that the desorption yield is more dependent on the surface treatment than on the base material for technical surfaces [19].

- The variation of the desorption yield with the nature of the incident ion shows an increase of the desorption yield with the ion mass which could be partly related to the enrichment of the surface with elements introduced in the substrate by the ion itself (e.g. C or O implanted by ions containing these elements as CO^+ or CO_2^+) [20].

Swift heavy ion-induced desorption phenomena are not treated in this context. Detailed information can be found in [21, 22].

The subject of this thesis is the measurement of the ion-induced desorption yield of technical materials such as OFHC-copper for various ions at different energies. Carried out in the vacuum group of the former AT department[2] at CERN its motivation originates from the special vacuum requirements of these materials which are exposed to the beam in the LHC.

In *Chapter 2* a general survey of CERN, its scientific achievements and the accelerator complex is given. An introduction to the LHC and its beam vacuum system together with the impact of beam related vacuum effects is given in *Chapter 3*. In *Chapter 4* the mechanism of ion-induced desorption together with the theoretical framework for the calculation of the desorption yield according to the so called "mass spectrometer method" and the calculation of the desorption cross section is presented. Further this chapter discusses the sensitivity of the measurement method of the desorption yield and deals with the energy loss of ions passing through matter. *Chapter 5* describes the experimental setup and its functionality. Modifications which were made on the pre-existing system in order to improve the measurement of the desorption yields are discussed. At the end of this chapter the measurement procedure is described. In *Chapter 6* the results of the ion-induced measurements are presented and discussed. A conclusion of the work is given in *Chapter 7*.

---

[2]meanwhile TE department

Chapter 2

CERN

2.1 Organization

The convention establishing CERN was signed on 29 September 1954 to re-establish fundamental physics research in post world war II Europe. From the original 12 signatories of this convention, membership has grown to the present 20 member states (cf. figure 2.1). The states' contributions to CERN for the year 2008 totalled in CHF 1.075.863 million (around $990 million) [23].

The acronym CERN — *Conseil Européenne pour la Recherche Nucléaire* (European Council for Nuclear Research), was given by a provisional council for setting up the laboratory in 1952. The acronym was retained, even though the name changed to the current *Organisation Européenne pour la Recherche Nucléaire* (European Organization for Nuclear Research) in 1954.

Currently it is the world's largest particle physics laboratory, situated in the northwest suburbs of Geneva on the border between France and Switzerland. Its main function is to provide the particle accelerators and other infrastructure needed for high-energy physics research. Numerous experiments have been constructed at CERN by international collaborations to make use of them and it currently has approximately 2600 full-time employees. Some 7931 scientists and engineers (representing 500 universities and 80 nationalities), about half of the world's particle physics community, work on experiments conducted at CERN.

2.2 Scientific achievements

The following list should point out several important achievements in particle physics which have been made during experiments at CERN. These include, but are not limited to:
- The discovery of neutral currents in the Gargamelle bubble chamber.
- The discovery of W and Z bosons in the UA1 and UA2 experiments.

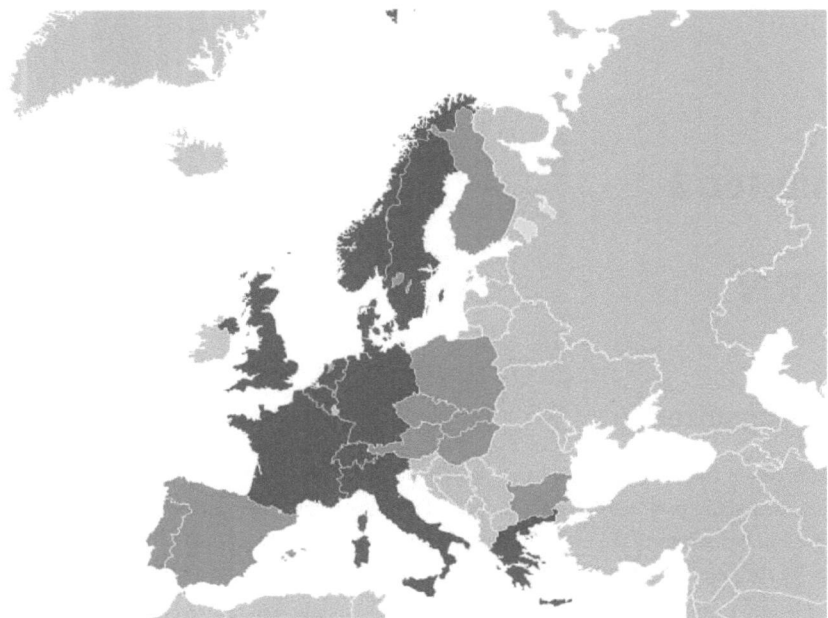

Figure 2.1: Member states of CERN [24] (blue: founding members; green: members who joined later).

- The determination of the number of neutrino families at the Large Electron Positron collider (LEP) operating on the Z boson peak.
- The first creation of antihydrogen atoms in the PS210 experiment.
- The discovery of direct CP-violation in the NA48 experiments.

The 1984 Nobel Prize in physics was awarded to Carlo Rubbia and Simon van der Meer for the developments that led to the discoveries of the W and Z bosons.

The 1992 Nobel Prize in physics was awarded to CERN staff researcher Georges Charpak for his invention and development of particle detectors, in particular the multiwire proportional chamber.

2.3 Accelerator complex

CERN operates a network of six accelerators and a decelerator (cf. figure 2.2). Each machine in the chain increases the energy of particle beams before delivering them to experiments or to the next more powerful accelerator (note: particles with a higher energy need a bigger magnetic field to keep

2.3. ACCELERATOR COMPLEX

them on their circular track. Since the ultimate magnetic field is limited, the only way to increase their energy is to increase the bending radius of the magnets).

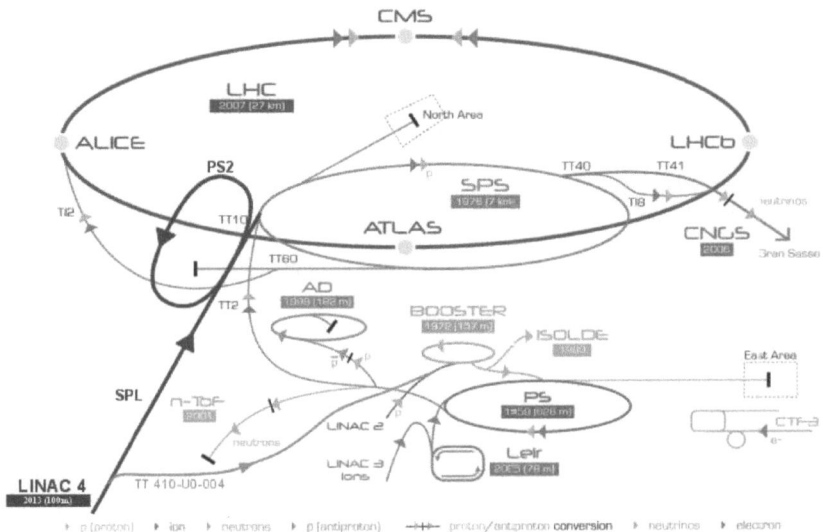

Figure 2.2: The accelerator complex at CERN.

Currently active machines are:

- Two linear accelerators generate low energy particles for injection into the Proton Synchrotron. The 50MeV LINAC2 is for protons, and the 4,2MeV/u LINAC3 is for heavy ions. A new linear accelerator (LINAC4) is under construction to replace the old LINAC2.

- The Proton Synchrotron Booster increases the energy of particles generated by the proton linear accelerator before they are transferred to the other accelerators.

- The Low Energy Ion Ring (LEIR) accelerates the ions from the ion linear accelerator, before transferring them to the Proton Synchrotron (PS). This accelerator was commissioned in 2005, after having been reconfigured from the previous Low Energy Anti-proton Ring (LEAR).

- The 28GeV Proton Synchrotron (PS), built in 1959 and still operating as a feeder to the more powerful SPS.

- The Super Proton Synchrotron (SPS), a circular accelerator with a diameter of 2 kilometers built in a tunnel, which started operation in 1976. It was designed to deliver an energy of 300GeV and was gradually upgraded to 450GeV. As well as having its own beamlines for fixed-target experiments, it has been operated as a proton-antiproton collider, and for accelerating high energy electrons and positrons which were injected into the Large Electron-Positron collider (LEP). From 2008 onwards, it will inject protons and heavy ions into the Large Hadron Collider (LHC).

- The On-Line Isotope Mass Separator (ISOLDE), which is used to study unstable nuclei. Particles are initially accelerated in the PS Booster before entering ISOLDE. It was first commissioned in 1967 and was rebuilt with major upgrades in 1974 and 1992.

- The Antiproton Decelerator (AD), which reduces the velocity of antiprotons to about 10% the speed of light for research into antimatter.

- The LHC which will be described in more detail in the following chapter.

Chapter 3

The LHC Project

The LHC[1], the *Large Hadron Collider*, has been designed to collide protons at a center-of-mass energy of 14TeV with a luminosity of 10^{34}cm$^{-2}\cdot$ s^{-1} [25, 26]. In addition it will provide collisions between lead nuclei up to a center-of-mass energy of 1150TeV[2].

The LHC was placed into the existing tunnel of the LEP collider whose operation had been stopped at the end of 2000 and which had been dismantled. Following this tunnel, the LHC has a circumference of about 26,7km.
As shown in figure 3.1 the general layout of the LHC has an eightfold structure, thereby following the layout of its predecessor LEP. Each of these octants consists of an *arc*, which basically contains the main bending magnets (the *main dipoles*), focussing and de-focussing (quadrupole) magnets and higher order correction magnets and is kept at cryogenic temperatures, followed by a so called *long straight section*[3] with a length of about 500m, which is kept at room temperature.

The two proton beams, each made up of 2835 tightly packed *bunches* of protons ($1,05 \times 10^{11}$ protons per bunch), resulting in a current of 0,536A per beam, will be circulating in clockwise and anti-clockwise direction in two separate beam pipes which are inserted into a common bending magnet. These beams are brought in collision at four points called *interaction points*.

At these four interaction points huge particle detectors will be installed to measure properties like energy and momentum of the particles emerging from the interaction point after the collision of high energetic particles. These detectors, depicted in figure 3.1, are:

- **A Toroidal LHC ApparatuS (ATLAS)** detector [27] consists of a series of ever-larger concentric cylinders around the interaction point where the proton beams from the LHC collide. It can be divided into four major parts: The *inner detector*, the *calorimeters*, the *muon spectrom-*

---
[1]The first two counter-rotating beams were successfully injected in the LHC on 10th of September 2008.
[2]This corresponds to $2 \times 82 \times 7$TeV, since a lead nucleus contains 82 protons. In reality, this means that each lead nucleus has only $82/208 \times 7 \approx 2,76$TeV/nucleon. This is due the extra mass of the neutrons in the nuclei.
[3]There are also so called *short straight sections* which are considered as a part of the arcs.

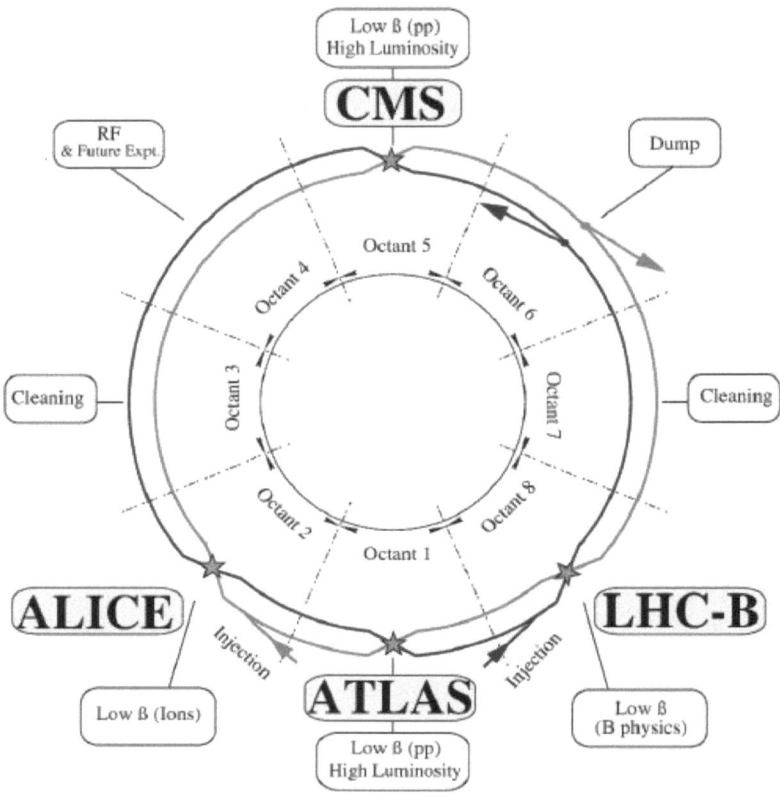

Figure 3.1: General layout of the Large Hadron Collider.

eter and the *magnet systems*. The magnet systems consist of an inner solenoid which produces a two tesla magnetic field surrounding the inner detector and of eight very large air-core superconducting barrel loops and two end-caps which are producing the outer toroidal magnetic field.

The main goal of ATLAS is the *search of the Higgs boson* [28] and the detector is designed to be sensitive to largest possible Higgs masses. The asymmetry between the behavior of matter and antimatter, known as *CP-violation* [28], will also be investigated and should explain according to the standard model [29] the lack of detectible antimatter in the universe. Furthermore due to the high energy and collision rates it should be possible to study the *top quark* [28], discovered at Fermilab in 1995, in more detail. New models of physics like the *broken symmetry* [28] can also be investigated.

- **Compact Muon Solenoid (CMS)** detector [30] is designed to meet the *same goals in physics like ATLAS* and will record similar sets of measurements on the created particles. In contrast it consists of *only one magnetic system* performed through a single superconducting solenoid which generates a magnetic field of 4T (about 100 000 times the magnetic field of the earth). The main detector systems are the *inner tracker* with ten layers of silicon strip detectors and silicon pixel detectors in the high occupancy range close to the interaction point, an *electro-magnetic calorimeter* with an excellent energy resolution and a *muon system* for momentum measurements up to highest luminosity.

- **A Large Ion Collider Experiment (ALICE)** [31] will be the dedicated detector to study the *strong interactions of matter* at extreme densities and high temperatures during heavy ion (e.g. lead) collisions. The data obtained will allow physicists to study a state of matter known as *quark-gluon plasma* [28], which is believed to have existed soon after the big bang. The detector consists of an *inner tracking system* made of six cylindrical layers of silicon detectors, *radiation and momentum detectors*, a *muon spectrometer* and an *electro-magnetic calorimeter*.

- **LHCb** experiment [32] should help us to understand why we live in a universe that appears to be composed almost entirely of matter, but no antimatter. Therefore it is designed to study the *CP-violation* in *B-mesons decay* [28]. The detector is a single arm forward spectrometer and consists of a *vertex detector, two ring imaging Cherenkov detectors* and of the *electromagnetic and hadronic calorimeters* which provide the measurement of the energy of electrons, photons and hadrons.

The *main dipoles*, LHC's main bending magnets, have to provide a nominal magnetic field of about 8,4T to accelerate protons to an unprecedented energy of 7TeV. The current required to create this field is about 11,8kA and constraints on geometry and heat budget require that the magnet coils are made of superconducting cables (cf. [33, 34]). These cables consist of fine strands (7μm diameter) of a Nb-Ti alloy which are twisted together and embedded in a copper matrix. Hence, the whole magnet has to be cooled below the critical temperature of the superconductor (the Nb-Ti alloy), which is achieved by means of superfluid helium at 1,9 K. About 80% of the total length of the accelerator will be held at these temperatures, thus making the LHC one of the biggest cryogenic facilities in the world [35].

As illustrated in figure 3.2, the beam pipes (inner diameter 50mm) for the two counter-rotating beams, together with a pair of superconducting coils each, are incorporated into a common iron yoke, thus being in direct contact with the cold mass and acting at the same time as the inner wall of the magnet cryostat[4] (the so called *cold bore*) which is a surrounding vacuum vessel to insulate the cold mass from ambient temperature [40, 41, 42, 43, 44]. This construction implies, that the walls of the beam vacuum system will have the same temperature as the cold mass itself, namely 1,9K during

---

[4]For a more detailed description cf. [36, 37, 38, 39]

operation. At this temperature, gases except Helium have a negligible vapor pressure, hence the beam pipe will effectively act as *cryopump* with basically unlimited capacity, making external pumping superfluous during operation [43, 44]. External pumps are required only for the initial pump-down of the vacuum system.

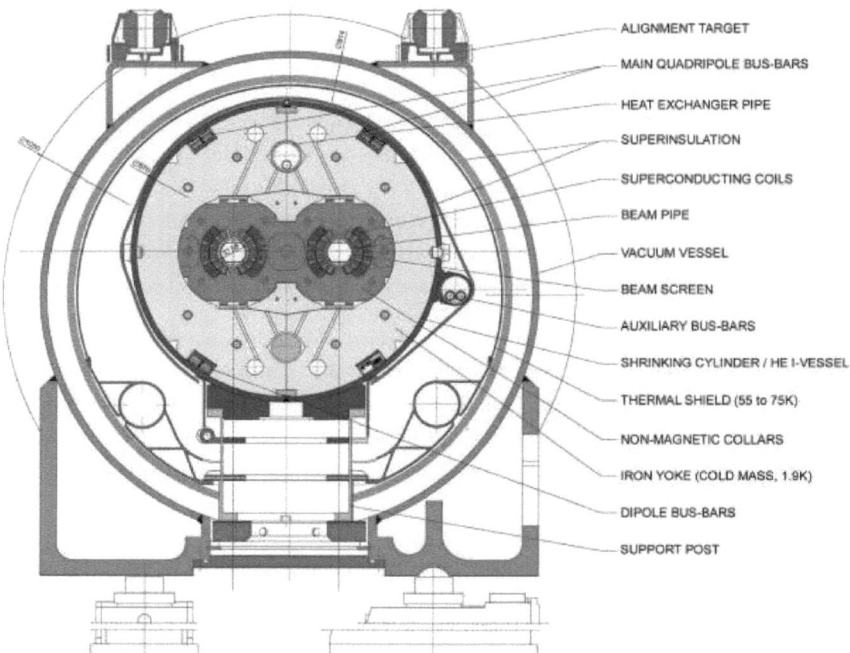

Figure 3.2: Cross section of a LHC dipole cryomagnet assembly.

The total length of an assembled dipole is about 16m and a total number of 1232 of these magnets will be built into the LHC. The smooth operation of the main dipoles is one of the crucial points in the operation of the LHC, since a *quench*, i. e. the transition from superconducting to resistive state, of a single magnet can interrupt machine operation for several weeks.

3.1 An introduction of the LHC beam vacuum system

The LHC has three different vacuum systems: the insulation vacuum for cryomagnets, the insulation vacuum for the helium distribution line and the beam vacuum. The requirements for the beam vacuum of the cold arcs and of the warm sections, e.g. the long straight sections, are different. To maintain a low residual gas pressure, as well as a low secondary electron emission yield (to avoid electron

3.1. AN INTRODUCTION OF THE LHC BEAM VACUUM SYSTEM

multipacting), the chambers in the warm sections are coated with a TiZrV non evaporable getter (NEG) which, after its activation, is a very good getter for H_2 and CO [45].

It will be the first superconducting accelerator which is exposed to intense synchrotron radiation. According to [46], the instantaneous power radiated by a charged particle, in this case a proton, traveling on a circular orbit is given by

$$P_{s.r.} = \frac{1}{4\pi\epsilon_0}\frac{2}{3}\frac{e^2}{c^3}\frac{v^4}{r^2}\gamma^4 \qquad (3.1)$$

where e is charge of the proton (i. e. the elementary charge), c the speed of light in vacuum, v the velocity ($v \approx c$), r the bending radius ($r = 2784,32$m for the main dipoles) and γ the relativistic factor ($\gamma = 7461$ for 7TeV protons). Substituting the numerical values into equation 3.1 results in a value for the instantaneous power radiated by one proton of $P_{s.r.} = 1,84 \times 10^{-11}$W. Having 2835 bunches with $1,05 \times 10^{11}$ protons per bunch distributed over the circumference of LHC (26658,883m) results in a average linear proton density of about $1,12 \times 10^{10}$m^{-1}, hence the linear heat load caused by synchrotron radiation in the main dipoles is about 0,2W· m^{-1}.

This heat load, if transferred to the cold mass, would increase excessively the heat dissipated in the superconducting magnets, hence the cold mass has to be shielded against synchrotron radiation. This is achieved by means of the so called *beam screen*, a racetrack shaped tube with two *cooling capillaries* attached to its two flat parts, which actively cool the beam screen to a temperature between 5K and 20K by means of pressurized Helium gas [43]. At this temperature all gases except hydrogen have a low enough vapor pressure to be condensed on the beam screen. The beam screen has a diameter of 44mm and the flat parts are separated by 36mm. A picture of a prototype beam screen, inserted into a sample beam pipe is shown in figure 3 3.

Another feature of the beam screen is a thin layer of oxygen free high conductivity copper (thickness about 50μm), which is co-laminated with the base material of the beam screen, a low permeability stainless steel. This layer is intended to carry the beam-induced image currents, thus reducing the machine impedance to an acceptable value. The chosen value for the thickness of the copper layer is in fact a compromise between low impedance and mechanical stability. Eddy currents, induced during a magnet quench are inversely proportional to the impedance. These eddy currents, in conjunction with the magnetic field of the dipole can result in very high Lorentz forces acting on and leading to deformation of the beam screen [47, 48, 49]. Without beam screen the image currents would flow through the beam pipe and produce an unacceptably high resistive heat load on the cold mass, which in turn would lead to a quench of the magnet. In the present configuration the heat load due to the image currents, about 0,1W· m^{-1} [50, 51], is intercepted by the beam screen.

A third important feature of the beam screen are the pumping slots incorporated in the flat parts of the beam screen, amounting to about 4% of the its total surface. Residual gas molecules can travel

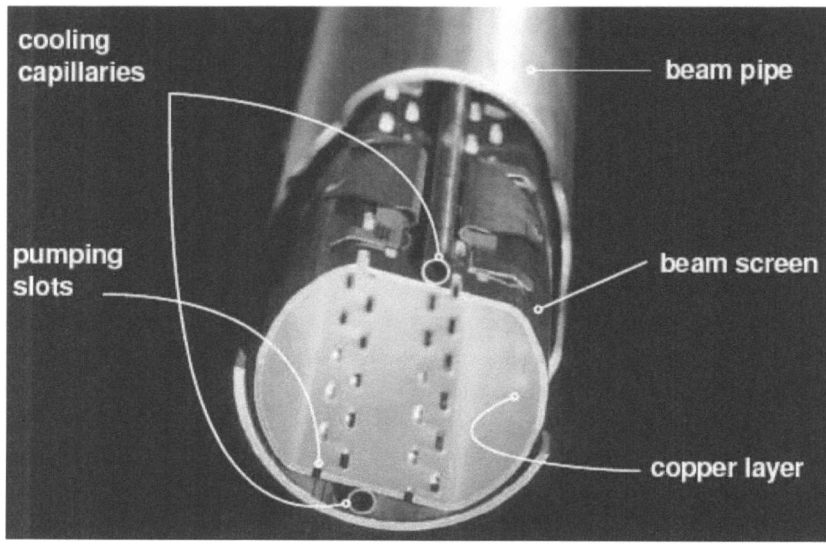

Figure 3.3: Picture of a prototype beam screen, inserted into a sample beam pipe.

through this pumping slots and reach the cold bore wall where they will permanently adsorbed [40]. The purpose of the pumping slots will be discussed in more detail in conjunction with the dynamic vacuum effects in the next section.

As a last, nevertheless important point, the beam loss due to nuclear scattering should be mentioned in the context of this introduction. A small fraction of scattered protons escapes from the aperture of the beam pipe and penetrates the surrounding material, thereby producing a shower of secondary particles which is finally absorbed by the cold mass. There is no way that these scattered particles can be absorbed by the beam screen and therefore the machine design includes an allowance of about $0,1 \text{W} \cdot \text{m}^{-1}$ for the linear heat load due to nuclear scattering for the two beams [44]. The linear heat load $P_{n.s.}$ (for one beam) can be expressed as

$$P_{n.s.} = \frac{I_{beam}}{e} N_G \, \sigma_{n.s.;G} \, E \qquad (3.2)$$

where $I_{beam} \approx 0,536\text{A}$ is the (nominal) current of the proton beam, e the elementary charge[5], N_G the number density of gas G, $\sigma_{n.s.;G}$ the cross section for nuclear scattering of a proton on a molecule of gas G (for Hydrogen molecules and 7TeV protons, it is $\sigma_{n.s.;H_2} \approx 5 \times 10^{-30} \text{m}^2$ [44]) and $E = 7\text{TeV}$ the proton energy. With the above mentioned numerical values and with $P_{n.s.} < 0,05 \text{W} \cdot \text{m}^{-1}$, it follows

---

[5] In fact, it is $I_{beam}/e = j$, i. e. the number of protons passing through an arbitrary cross section of the beam pipe per unit time.

from equation 3.2 that $N_{H_2} \leq 2,66 \times 10^{15} \text{m}^{-3}$ and in consistency with this requirement an upper limit for the residual number density of Hydrogen molecules of $N_{H_2} = 1 \times 10^{15} \text{m}^{-3}$ has been chosen for the design of the beam vacuum system to ensure a beam lifetime of 100 hours [44]. Assuming a gas temperature of 10K, the resulting upper limit for the partial pressure of Hydrogen is given as $p_{H_2} = 1,38 \times 10^{-7} \text{Pa} \approx 1 \times 10^{-9} \text{Torr}$. Corresponding values for other gases can be found in [43] or [44].

3.2 Beam related dynamic vacuum effects and their impact on LHC

3.2.1 Desorption by synchroton radiation

Synchrotron radiation photons hit the inner surface of the beam screen where they are either absorbed or scattered, thereby releasing all or part of their energy. In consequence, residual gas molecules adsorbed on this surface can be released into the gas phase, i. e. they are *desorbed*, if the available energy exceeds the energy of the bond between the molecule and surface.

In general, the energy spectrum of the synchrotron radiation, emitted by a charged particle moving on a circular orbit with almost speed of light can be characterized by the so called *critical energy*, given by [46]

$$E_c = \frac{3}{2}\frac{\hbar c}{r}\gamma^3 \qquad (3.3)$$

and the number of photons emitted by each beam particle per unit time can be calculated from [46]

$$j_\gamma = \frac{15\sqrt{3}}{8}\frac{P_{s.r.}}{E_c} = \frac{5\sqrt{3}}{6}\frac{\alpha c}{r}\gamma \qquad (3.4)$$

where $\hbar$ is the reduced Planck constant, c the speed of light in vacuum, α the fine-structure constant, r the bending radius and γ the relativistic factor. In the case of LHC, it is $E_c \approx 44,1 \text{eV}$ and $j_\gamma \approx 8,46 \times 10^6 \text{s}^{-1}$. With an average linear proton density of $\mathcal{N}_p \approx 1,12 \times 10^{10} \text{m}^{-1}$ an average *linear photon flux* can be calculated as

$$\dot{\mathcal{N}}_\gamma = j_\gamma \mathcal{N}_p \approx 9,45 \times 10^{16} \text{s}^{-1} \cdot \text{m}^{-1} \qquad (3.5)$$

The number of desorbed molecules is proportional to this photon flux, hence

$$\dot{\mathcal{N}}_G = \eta \dot{\mathcal{N}}_\gamma \qquad (3.6)$$

The constant of proportionality η is called the *molecular desorption yield* and is usually given in units of molecule · photon^{-1} [6]. η is in fact not a constant but depends on several factors, among others the *nature of the desorbed gas*, the *surface material, temperature* and *pre-treatment* and the *"history" of the surface*, i. e. the number of photons the surface has been exposed to.

The phenomenon of desorption by synchrotron radiation, also called *photon stimulated desorption* or *photon induced desorption*, has been subject to extensive studies in the past (cf. [52, 53] in the context of LEP, [54, 55, 56, 57, 58, 59, 60] in the context of LHC and/or the SSC, the Superconducting Super Collider). Values for η at conditions relevant for LHC, given in above cited literature, range from some 10^{-3} molecule · photon^{-1} for H_2 to some 10^{-5} molecule · photon^{-1} for CO_2 and CH_4. After a long exposure of the surface to synchrotron radiation these values are reduced by 1 ... 2 orders of magnitude, an effect which is also called *beam scrubbing* or *beam cleaning*.

In the case of cryogenic vacuum systems, molecules in the gas phase are readily pumped by the cold walls. These molecules are only lightly bound to the surface, i. e. *physisorbed* and they can be re-desorbed by synchrotron radiation with a much higher yield. This process is called the *recycling* of previously physisorbed molecules and the corresponding desorption yield, usually denoted as η', can exceed η by several orders of magnitude [55, 59]. Recently, the cracking of adsorbed molecules by synchrotron radiation has been identified as an additional mechanism which can contribute significantly to the gas load in the beam vacuum system [60].

At this point the importance of the pumping slots of the beam screen can be well explained. A fraction of the desorbed and recycled gas molecules can travel through these slots and reach the surface of the cold bore, Since this surface is shielded from synchrotron radiation, these molecules are not recycled and hence can be permanently cryosorbed on the beam pipe. Thus, unlike the inner surface of the beam screen where molecules are continuously recycled, the pumping slots provide the means to remove gas effectively and permanently from the beam vacuum system.

3.2.2 Ion-induced pressure instability

Positive ions can be produced in the beam vacuum system through the ionization of residual gas molecules by the beam particles with typical ionization cross sections for 7TeV protons in the range of 10^{-22}m^2 [61]. These positive ions are then repelled by the positive space charge of the beam and accelerated towards the beam screen where they transfer their kinetic energy onto the surface. In the arcs of the LHC ion energies at impact are typically in the range of several 100eV [44, 62, 63]. Like with photon stimulated desorption, the number of molecules, desorbed due to the impact of energetic

---

[6]Nevertheless the physical unit of η is 1.

3.2. BEAM RELATED DYNAMIC VACUUM EFFECTS AND THEIR IMPACT ON LHC

ions, is proportional to the number of incident ions, hence

$$\dot{\mathcal{N}}_G = \eta_i \dot{\mathcal{N}}_+ \tag{3.7}$$

with $\dot{\mathcal{N}}_G$ being the linear flux of molecules of the species G, desorbed from the beam screen and $\dot{\mathcal{N}}_+$ the linear flux of ions hitting the beam screen. η_i is like before called the *molecular desorption yield* but this time expressed in units of molecule · ion^{-1}. Again, η_i is not a constant. It depends not only of the *nature of ions and desorbed molecules*, the *ion energy, nature and temperature of the surface* [64, 65], but also on the *surface preparation and condition* [9]. In the case of cryogenic vacuum systems, we can again distinguish between tightly bound, chemisorbed molecules and physisorbed molecules. Whereas it is $\eta_i \approx 1 \dots 10$ molecule · ion^{-1} in the case of chemisorbed molecules it can be several thousand molecules per ion for physisorbed molecules [66, 67].

If one gas species G is dominant, $\dot{\mathcal{N}}_+$ can be expressed as

$$\dot{\mathcal{N}}_+ = \sigma_{i;G}\, N_G\, \frac{I_{beam}}{e} \tag{3.8}$$

where $\sigma_{i;G}$ is the ionization cross section of this species, N_G its number density, I_{beam} the proton beam current and e the elementary charge.

It can be seen from the preceding paragraphs (N.B. equations 3.7 and 3.8) that the process of ion stimulated desorption in a beam vacuum system is "self-amplifying"[7] and could result in a pressure run-away (or *pressure instability*) if the gas is not pumped away with a sufficient pumping speed. This effect has been observed at the ISR at CERN [61]. However, in the cold parts of the LHC the ion-induced desorption should not pose any serious problems to the beam vacuum due to the distributed cryo-pumping of the cold walls [44].

3.2.3 Beam induced electron multipacting

Electron multipacting is a phenomenon known from high power radio frequency and microwave cavities where it manifests itself in RF power consumption and break down. It is caused by the synchronous motion of free electrons in an alternating electric field. First free electrons are produced by field emission, photo-electric effect, or ionization of residual gas molecules by cosmic rays. These electrons are accelerated towards the surface of the cavity by the electric field where they, when hitting the surface with sufficient energy, can produce secondary electrons. If the electric field changes its direction at the same time, these secondary electrons are accelerated towards the opposite surface where they in turn produce additional electrons. If the *secondary electron yield*, i. e. the number of secondary electrons produced per incident electron, exceeds unity, the number of electrons which are "bouncing" back and forth is increasing exponentially (also known as the "build up of the *electron*

---

[7]hence an increase of gas density results in an increase of the ion flux which in turn results in an increase of the molecular desorption rate which results in an even faster increase of the gas density ...

15

cloud"), finally leading to the break-down of the cavity [2, 68, 69]. In general, for electron multipacting to develop, the following two condition must be fulfilled [70]:

- The electron must be able to traverse the vacuum chamber in synchronism with the electric field and

- the electron energy at impact must result in an secondary electron yield greater then unity.

Since several years a similar effect has been observed in the beam pipes of high current proton accelerators (cf. [70, 71, 72, 73]). This phenomenon is called *beam induced multipacting* because the alternating electric field is generated by the bunched proton beam. In the case of LHC with its cryogenic vacuum system, the build-up of an electron cloud can have the following implications:

- Excessive heat load on the vacuum chamber surfaces (computer simulations give values up to $15 \text{W} \cdot \text{m}^{-1}$ [74, 75]),

- strong pressure rise due to the desorption of adsorbed molecules from the beam screen surface by impact of electrons (*electron stimulated desorption*) and

- coherent oscillations of the proton beam with the electron cloud, leading to emittance growth and luminosity decrease or even beam loss [40].

In the arcs of the LHC, primary electrons are massively created through the photo-electric effect due to the high flux of synchrotron radiation photons. The production rate of photo-electrons per proton (j_{e^-}) is proportional to the production rate of synchrotron radiation photons (j_γ):

$$j_{e^-} = 0,45 \, j_\gamma \, Y \tag{3.9}$$

Y is the *effective quantum yield*, i. e. the number of photo-electrons produced per incident photon. A value of $Y \approx 0,1$ is commonly assumed for LHC relevant conditions [76]. Only about 45% of the incident photons have enough energy[8] to produce photo-electrons [75], hence the factor of 0.45 in equation 3.9.

Synchrotron radiation photons emitted by a traveling proton bunch hit the circumference of an arbitrary cross section (normal to the beam axis) of the beam screen at about the same moment as the proton bunch travels through this cross section, hence the instantaneous production rate of photo-electrons varies in synchronism with the bunch structure of the beam [76]. Furthermore, since photons are preferably emitted in the forward tangential direction of the beam orbit, the instantaneous production rate of photo-electrons has also an azimuthal dependency. For a surface material with high reflectivity, photons are likely to be reflected many times before producing a photoelectron, hence the photo-electrons are distributed uniformly over the beam screen surface. On the other

---

[8] i. e. an energy greater then the work function of copper, about 4eV [74].

hand, if the reflectivity of the surface is low, synchrotron radiation photons are likely to produce photo-electrons already at their first impact on the surface, hence the photo-electron distribution follows that of the photons. The initial electron distribution has an influence on the development of beam induced multipacting especially in the parts of the accelerator where a strong magnetic field is present, e. g. in the main dipoles[9].

Photo-electrons receive approximately a "kick-like" acceleration towards the beam axis by the passing proton bunch, which is proportional to the number of protons in the bunch and hence proportional to the beam current. From the condition for the onset of multipacting – electrons must be able to traverse the beam vacuum from wall to wall before the arrival of the next bunch – a critical beam current can be calculated. For nominal LHC parameters the energy gain during kick acceleration is about 200eV and the critical beam current for the onset of multipacting is $I_{crit} \approx 0,19$A [70].

As with the RF related multipacting, the secondary electron yield δ must exceed unity to develop beam induced multipacting. In fact, since some of the secondary electrons can get out of phase with the electron cloud movement and are lost for further multiplication, the critical value of the secondary electron yield $\delta_{crit.}$ is greater than 1. For nominal LHC operating conditions it is $\delta_{crit.} \approx 1,3$ [40].

Because of the critical influence of the secondary electron yield for the development for multipacting and hence for the operation of LHC, this parameter has been the topic of extensive research work carried out in the LHC vacuum group over the last years (cf. [77, 78, 79, 80, 81]). A summery of the main experimental results concerning the secondary electron yield of copper is given in [77], and the curves shown in figure 3.4 show typical examples for these results. The curve denoted as "as received" refers to a surface prepared for installation in the vacuum system, whereas the "fully conditioned" curve refers to a surface with all contaminants stripped off. It can be seen that the maximum secondary electron yield can be reduced significantly by means of proper conditioning of the surface, i. e. during beam operation by beam scrubbing or by surface conditioning (bake-out, argon glow discharge treatment, ...).

As a last point in this section the desorption of gas molecules by electrons, called *electron stimulated desorption*, should be mentioned. The desorption rate is proportional to the rate of impinging electrons and characterized by the *molecular desorption yield* η_e (given in units molecule · electron^{-1}, cf. section 3.2.1 and 3.2.2)

$$\dot{\mathcal{N}}_G = \eta_e \dot{\mathcal{N}}_{e^-} \qquad (3.10)$$

Again, η_e is not a constant but depends on the *energy of impinging electrons*, the *nature of the desorbed gas*, the *material, temperature, treatment and history of the surface* (cf. [64, 65, 82, 83, 84]).

---
[9]Due to Lorentz forces, electrons are bound to move in spirals around the magnetic field lines.

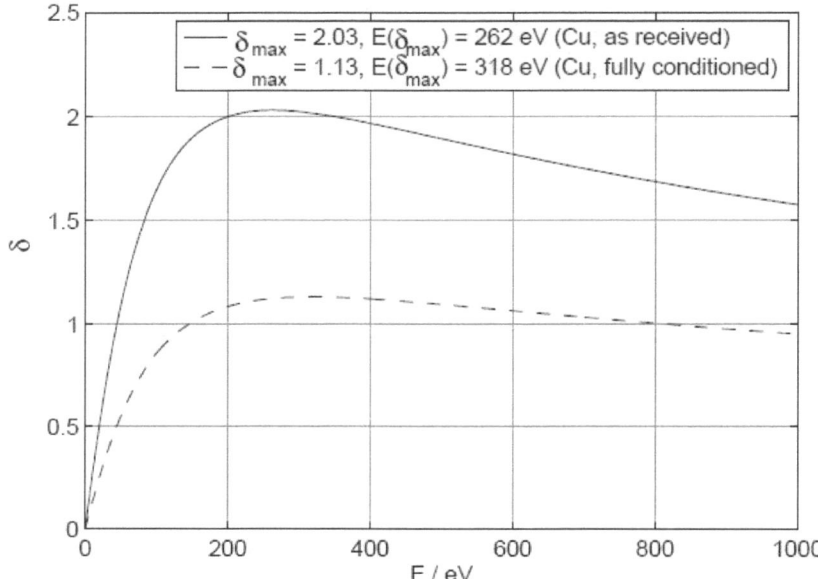

Figure 3.4: Typical plots of the secondary electron yield of copper as a function of the primary electron energy. Data taken from [77].

Due to the electron stimulated desorption, the build-up of an electron cloud manifests itself also by a strong increase in the residual gas pressure. In fact, pressure rises up to a factor of 60 could be observed in the SPS[10] during runs with LHC-type beams [72]. Thus, an electron cloud in the beam vacuum system of the LHC would most probably raise the pressure during the conditioning period to an unacceptably high level for the cryogenic system of the LHC.

These beam related dynamic vacuum effects and their impact on LHC can be simulated with computer models like the VASCO (VAcuum Stability COde) code [85].

---

[10] Super Proton Synchrotron, one of CERN's accelerators which will be used to pre-accelerate and inject beam particles into the LHC.

Chapter 4

Theoretical Framework

4.1 Ion-induced desorption mechanism

Ions which are created in accelerators due to the ionization of residual gas atoms by the circulating beam have typically energies in the range of eV to keV [62, 63, 66]. For these slow ions, according to the theory of binary collisions of Winters and Sigmund [5], the following three different mechanisms for the ion-induced desorption process of adsorbed atoms/molecules can be considered [86, 87]:

- A projectile ion hits an adsorbate atom at the topmost layer which might be reflected directly or, at high enough energy, after some penetration into the substrate (mechanism 1 in figure 4.1(a)).

- A projectile ion penetrates into the substrate, but is eventually reflected. On its way out it may knock off an adsorbate atom at the topmost layer (mechanism 2 in figure 4.1(b)).

- The projectile ion causes an outward flux of substrate atoms in direction to the surface. Some of these atoms may knock off adsorbate atoms at the topmost layer on their way out (mechanism 3A in figure 4.1(c) and mechanism 3B in figure 4.1(d)).

This model is well adapted to study the desorption of *chemisorbed gases* from metals (i.e. binding energies in the eV range) and a number of simulations, based on this model, have been developed [88, 89, 90]. However, in the case of *physisorbed gases* with binding energies in the meV range, e.g. the desorption of condensed gases, other desorption processes like electronic sputtering can be predominant [91].

For swift heavy ions (with energies of MeV or even higher) which are often partially or totally stripped, the situation is completely different. In this case desorption yields can be predicted by the help of thermal spike models (cf. [21]).

CHAPTER 4. THEORETICAL FRAMEWORK

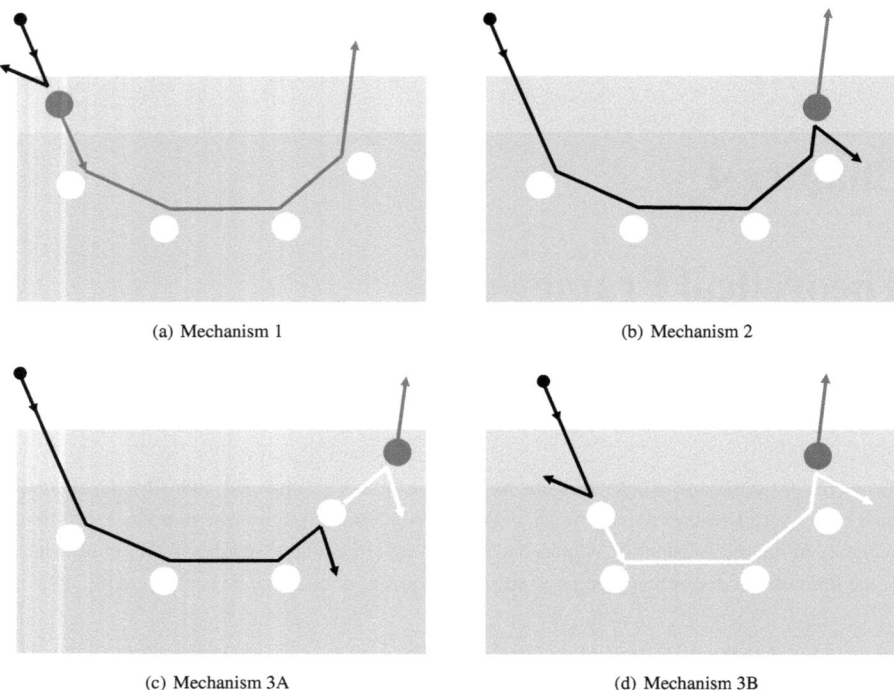

Figure 4.1: Schematic representation of the ion-induced desorption mechanisms, where the black circle represents the projectile ion, the red circle an adsorbate atom, and the white circle a substrate atom.

4.2 Calculation of the desorption yield

By the help of the ideal gas law it is possible to calculate the number of particles N inside a system at a given volume V and temperature T ($k_B = 1,38 \times 10^{-23}$ J/K is the so called Boltzmann-constant).

$$N = \frac{P \cdot V}{k_B \cdot T} \tag{4.1}$$

A target bombarded by particles within this system leads to a pressure rise ΔP. The number of particles which are desorbing from the target can be calculated by

$$N_{des} = \frac{\Delta P \cdot V}{k_B \cdot T} \tag{4.2}$$

This number divided by the number of impinging ions N_{ion} on the target leads to the so called des-

4.2. CALCULATION OF THE DESORPTION YIELD

orption coefficient η.

$$\eta = \frac{N_{des}}{N_{ion}} = \frac{\Delta P \cdot V}{N_{ion} \cdot k_B \cdot T} \tag{4.3}$$

This is correct as long as the volume flow rate of the pumping system (the pumping speed S) is neglected. As soon as S is greater than zero, the first pressure drop is followed by an exponential pressure decrease and its time response is depending on the pumping speed in the following way:

$$P\frac{dV}{dt} + V\frac{dP}{dt} = 0 \tag{4.4}$$

For a given pressure the changing of the volume with time (dV/dt) is equal to the pumping speed S of the system

$$P \cdot S = -V\frac{dP}{dt} \tag{4.5}$$

hence

$$t = \frac{V}{S} \cdot \ln \frac{P_0}{P_f} \tag{4.6}$$

where P_0 is the pressure right after bombardment and P_f the pressure in the system after the time t.

In a real vacuum system the outgassing rate of the system walls Q_{stat} is in a steady state with the pumping speed given by the pump. During desorption the pressure rises and after a while a new, higher steady state pressure is reached which is more or less constant with the time. To calculate the desorption coefficient under these circumstances equation 4.3 has to be differentiated due to time

$$\eta = \frac{\Delta P \cdot \dot{V}}{\dot{N} \cdot k_B \cdot T} = \frac{\Delta P \cdot S}{j \cdot k_B \cdot T} \tag{4.7}$$

where $\dot{N} = j$ is the flux of the projectiles impinging on the target, hence given as projectiles per time and could be expressed by $j = \frac{I_{ion}}{e}$ with the elementary charge e.
η is therefore given as:

$$\eta = K \cdot \frac{S}{I_{ion}} \cdot \frac{I_{RGA}}{\varsigma_{RGA} \cdot i_e} \tag{4.8}$$

where the constant $K = e/k_B T$, I_{RGA} is the ion current measured with the Residual Gas Analyzer (RGA), ς_{RGA} is a gas dependant sensitivity factor of the RGA and i_e is the emission current of the RGA.

Before the desorbed gas is detected in the gauge it can hit the chamber walls for several times with a sticking coefficient greater than zero which can differ for several experimental systems. Therefore the desorption coefficient is called as an *effective desorption coefficient* and its values could be compared only for several samples of one experiment but not for different experimental setups [21].

4.3 Calculation of the desorption cross section

The ion-induced desorption can be expressed using the following formulas, as was proved indirectly by experiments [5]:

$$N_d(t) = N_0 \cdot e^{-J\sigma t} \quad (4.9)$$

where $N_d(t)$ is the number of particles on/in the solid available for desorption per unit surface area at time t, N_0 the total number of particles on/in the solid available for desorption per unit surface area at time t_0, J the flux density of incident ions, and σ the desorption cross-section.
From equation 4.9 the following deduction can be made:

$$\frac{dN_d(t)}{dt} = -J\sigma N_0 \cdot e^{-J\sigma t} = -J\sigma N_d(t) \quad (4.10)$$

Assuming that

$$\eta(t) = -\frac{1}{J} \cdot \frac{dN_d(t)}{dt} \quad (4.11)$$

leads to the simple correlation between the desorption cross-section σ and the desorption coefficient $\eta(t)$:

$$\eta(t) = \sigma \cdot N_d(t) \quad (4.12)$$

It is clear from equation 4.12 that η is linear with $N_d(t)$ and the slope of this line is equal to σ under the condition, that σ is independent of $N_d(t)$.
In the case of room-temperature technical surfaces, the number of particles per unit area is not known and the cross-section can be obtained from desorption measurements using equation 4.12 and the following relation

$$Q_d(t) = N_0 - N_d(t) \quad (4.13)$$

where $Q_d(t)$ is the total number of desorbed molecules per unit surface area at time t. Hence

$$Q_d(t) = N_0 - \frac{\eta(t)}{\sigma} \quad (4.14)$$

and

$$\eta(t) = \sigma \cdot [N_0 - Q_d(t)] \quad (4.15)$$

Plotting η as a function of the number of desorbed molecules $Q_d(t)$ yields a straight line which slope represents the desorption cross-section and the intercept with the ordinate represents the product of

N_0 and σ [86, 92].

4.4 Minimum measurable desorption yield

The *sensitivity of the measurement method* can be defined as the *minimum η* which can be measured [86].

$$\eta_{min} = K \cdot \frac{\Delta P_{min}}{I_{max}} \cdot S \tag{4.16}$$

ΔP_{min} is the *minimum measurable pressure variation* (fixed arbitrarily to 1/100 of the base pressure which is a small value, difficult to achieve)

$$\Delta P_{min} = \frac{1}{100} \cdot P = \frac{1}{100} \cdot \frac{Q_{stat}}{S} \tag{4.17}$$

where S is the pumping speed of the measuring system, Q_{stat} the total outgassing of the measuring system and I_{max} the maximum (electron/ion) current available for desorption.
Hence η_{min} given by

$$\eta_{min} = \frac{K}{100} \cdot \frac{Q_{stat}}{I_{max}} \tag{4.18}$$

is independent of the pumping speed and only determined by the available current and the static outgassing of the measuring equipment (cf. figure 4.2).

However, there is a *limit to the measurable current* I_{min} given by the current amplifiers and hence to ΔP_{min} independent of the static pressure in the system. If one consider the minimum current I_{min} fixed by the current amplifier to be 10^{-11}A, then another limit is set to η_{min}:

$$\eta_{min} \geq K \cdot \frac{S}{I_{max}} \cdot \frac{I_{min}}{\varsigma_{RGA} \cdot i_e} \tag{4.19}$$

In this case, η_{min} is directly proportional to the pumping speed and inversely proportional to the RGA sensitivity ς_{RGA}. Hence an increasing of the pumping speed is a limitation for the measurement of a small desorption yield.

Figure 4.2 summarizes the various limitations to η_{min} (for $\varsigma_{RGA} = 10^5 \text{Pa}^{-1}$, $I_{min} = 10^{-11}$A and $I_{max} = 10^{-7}$A).

4.5 Energy loss of ions passaging through matter

During the bombardment of solids by ions, electrons or photons a lot of phenomena take place which are closely connected. These phenomena include the *energy loss* of charged particles in solids, *energy deposition* and the *ejection of secondary particles* during bombardment with charged particles. While

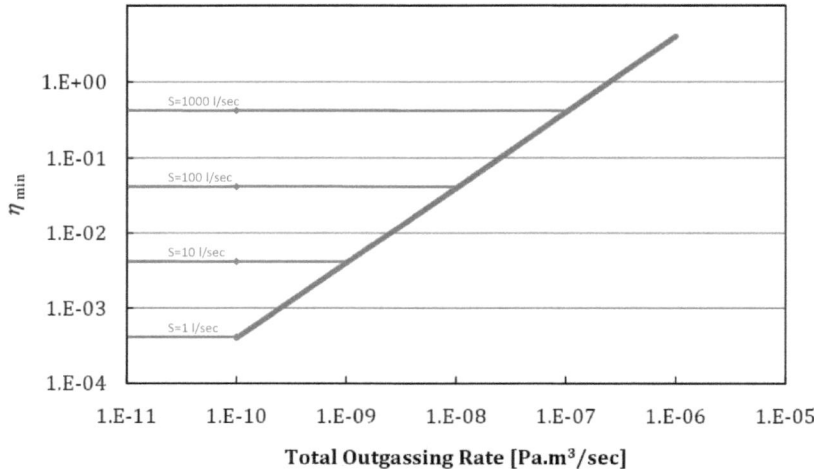

Figure 4.2: Various limitations to η_{min}: A decrease of η_{min} with the system outgassing rate down to a limitation is only depending on the pumping speed.

the term *desorption* refers usually to the removal of less than a monolayer by electronic transitions at the very surface, *sputtering* usually comprises particles ejected as a result of momentum transfer to target particles or electronic transitions.

The key quantity in energy loss considerations is the *stopping force* dE/dx (which in the past was named stopping power [93]). It can be considered as the force which the medium exerts on the penetrating particle:

$$dE/dx = N \cdot S(E) \quad (4.20)$$

where N is the number density of atoms in the medium and $S(E)$ the *stopping cross section*, in which the dependence on the kinetic energy E of the primary particle is explicitly written. The *stopping cross section is a density independent quantity except for ultra-relativistic energies* and has the dimension of energy times area. Note that in many tables authors do not distinguish between stopping cross section and stopping force with the dimension of energy per length [93].

The collisions between the primary ions and the atoms in a solid can be divided into collisions between the primary particle and the nuclei and those between the primary and the electrons. The first collisions take place for small impact parameters and lead to a large angle scattering process, whereas the latter ones leads to energy loss without any significant deflection of the primary particle. The

4.5. ENERGY LOSS OF IONS PASSAGING THROUGH MATTER

stopping cross section $S(E)$ can be split up into

$$S(E) = S_n(E) + S_e(E) \qquad (4.21)$$

where $S_n(E)$ is the *nuclear stopping cross section* and $S_e(E)$ the *electronic stopping cross section*. The term 'nuclear' is misleading, since the primary ion interacts with the screened nuclei rather than the bare nuclei [93]. A theoretical framework for a general treatment of energy loss to the nuclei as well as to the electrons can be found in [94, 95].

By taking the ratio $\xi(E) = S_e/S_n$ as a measure of the division of energy dissipation into electronic and atomic motion, figure 4.3 shows that there is a natural division into three energy regimes of different behavior as long as $Z_1 \cong Z_2$ [94]:

- At low energies in *regime I* the nuclear stopping is dominating and relatively little energy goes into electronic motion. The upper bound of *regime I* is given by E_c, which is described later on.

- Above this energy in *regime II* the nuclear stopping falls off, while the electronic stopping goes on increasing as $E^{1/2}$. Hence the ratio ξ increases rapidly and the fraction of energy going into electronic motion must increase correspondingly. The upper bound of *regime II* is given by E_1.

- Finally in *regime III* for energies above E_1 the electronic stopping starts decreasing monotonously, and the ratio ξ, though still increasing, approaches a maximum value.

4.5.1 Nuclear energy loss

When an ion with the nuclear charge $Z_1 \cdot e$ (e is the elementary charge) approaches to a surface atom with the nuclear charge $Z_2 \cdot e$ along a radius vector r, the ion will be scattered by the Coulomb repulsive interaction of the two particles. However, only for very high energies (MeV) the collision can be described by a pure Coulomb force because with increasing velocity the projectile loses the electrons and is at high velocities completely stripped. At intermediate velocities (*regime II* in figure 4.3) the projectile electrons with velocities exceeding the projectile velocity v will stick to the projectile, while the slower ones, belonging to the outer shells, will be stripped. However, regardless of the initial ion charge state, which may be very far from the equilibrium charge state, an ion beam will approach charge state equilibrium after having penetrated a few layers from the surface [93].

Due to the screening from the electrons at lower energy regimes (keV) it has to be considered that the interaction potential falls off faster than 1/r. The most widely used potentials may be considered as a Coulombic term (1/r) multiplied by a screening function. The Coulombic term arises from the repulsive interaction between the two positive point charges while the screening function models the

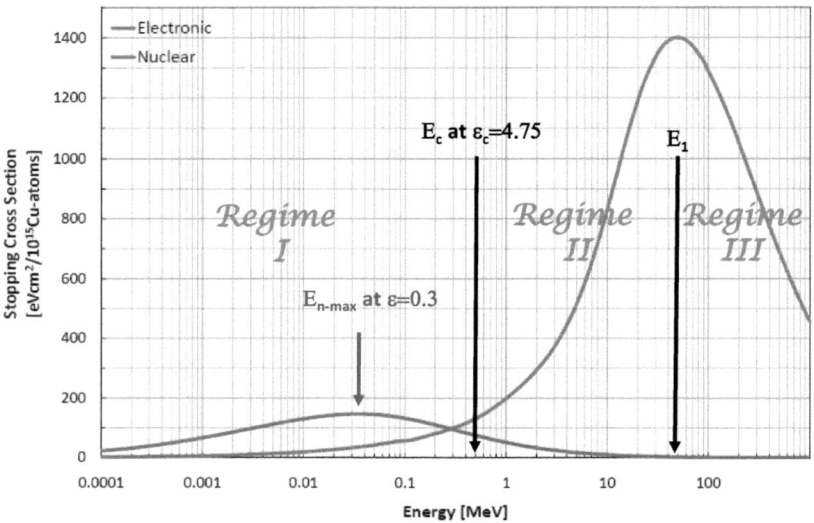

Figure 4.3: Electronic and nuclear stopping cross sections for Ar^+-ions incident on copper.

influence of the surrounding electron clouds [96, 97]. Such a screened Coulomb potential can be written in the form

$$V(r) = \left(\frac{Z_1 \cdot Z_2 \cdot e^2}{r}\right) \cdot \phi(r) \tag{4.22}$$

The screening function is given by

$$\phi(r) = \sum_{i=1}^{3} C_i e^{-b_i r/a} \tag{4.23}$$

where a is Lindhard's screening length given by

$$a = \left(\frac{9\pi^2}{128}\right)^{1/3} \cdot \frac{a_0}{(Z_1^{2/3} + Z_2^{2/3})^{1/2}} \tag{4.24}$$

with a_0 the Bohr radius

$$a_0 = \frac{4\pi\epsilon_0 \hbar^2}{m_e e^2} \approx 5,29 \cdot 10^{-11} m \tag{4.25}$$

The parameters C_i and b_i for the screening function can be found in [98].

The exact classical solution of the equation of motion of a particle in such a central-force potential,

4.5. ENERGY LOSS OF IONS PASSAGING THROUGH MATTER

$V(r)$, results in a scattering angle, θ, in the center-of-mass system given by

$$\theta = \pi - 2 \int_{r_0}^{\infty} \frac{p\,dr}{r^2[1 - V(r)/E_i - p^2/r^2]^{1/2}} \tag{4.26}$$

where p is the impact parameter, r_0 is the turning point (distance of closest approach) – given by the root of the expression in the square root in equation 4.26 and E_i is the energy of the particle in the center-of-mass system. E_i is related to the initial kinetic energy E by

$$E_i = M_2 E/(M_1 + M_2) \tag{4.27}$$

where M_1 and M_2 are the incident- and target-atom masses, respectively. Differentiation of equation 4.26 leads to the differential scattering cross section $\sigma(\theta)$, given by

$$\sigma(\theta) = \frac{-p}{\sin\theta} \cdot \frac{dp}{d\theta} \tag{4.28}$$

The scattering cross section is related to the energy transfer cross section $\sigma(T)$ by

$$\sigma(T) = (4\pi/\gamma E) \cdot \sigma(\theta) \tag{4.29}$$

where $\gamma = 4M_1 M_2/(M_1 + M_2)^2$ and T is the energy transferred to the target atom

$$T = \gamma E \sin^2 \frac{\theta}{2} \tag{4.30}$$

Finally the nuclear stopping force can be obtained from

$$\left(\frac{dE}{dx}\right)_n = N \cdot \underbrace{\int_0^{T_m} T\sigma(T)dT}_{S_n(E)} \tag{4.31}$$

were T_m is the maximum transferred energy ($T_m = \gamma E$).
By the help of the reduced energy ε

$$\varepsilon = \frac{aM_2 E}{Z_1 Z_2 e^2 (M_1 + M_2)} \tag{4.32}$$

Lindhard and coauthors [99] showed that the nuclear stopping force for all target-beam combinations could be expressed in terms of the reduced energy as

$$\left(\frac{dE}{dx}\right)_n = \frac{\pi a^2 \gamma N}{(\varepsilon/E)} S_n(\varepsilon) \tag{4.33}$$

where $S_n(\varepsilon)$ is given by [100, 101]

$$S_n(\varepsilon) = \frac{3,441\,\varepsilon^{1/2} \ln(\varepsilon + 2,718)}{1 + 6,355\,\varepsilon^{1/2} + \varepsilon(6,882\,\varepsilon^{1/2} - 1,708)} \tag{4.34}$$

The absolute magnitude of the nuclear stopping force is determined partly by the factor (ε/E) in the denominator, which means that *for heavy atoms on a heavy target* the factor becomes small and, in turn, leads to a *large nuclear stopping*. Equation 4.32 also demonstrates that ε decreases with increasing atomic number (and mass) of the projectile. The *maximum of the nuclear stopping force*, which for the classical Thomas-Fermi model occurred at $\varepsilon_{n-max} \cong 0,3$, is therefore shifted to higher energies for heavy projectiles [93, 96, 98].

4.5.2 Electronic energy loss

The electronic stopping in the low and intermediate energy regime (I and II in figure 4.3) can be determined by Lindhard-Scharff treatment [102].
Within a considerable velocity interval, i.e. for $v < v_1 \cong v_0 \cdot Z_1^{2/3}$ ($v_0 = e^2/\hbar$ is the Bohr velocity), the *electronic stopping cross section is nearly proportional to* v and is of the order of

$$S_e(E) \cong 8\pi e^2 a_0 \frac{Z_1^{7/6} Z_2}{\left(Z_1^{2/3} + Z_2^{2/3}\right)^{3/2}} \cdot \frac{v}{v_0} \tag{4.35}$$

In reduced units it can be written as

$$S_e(\varepsilon) = k \cdot \varepsilon^{1/2} \tag{4.36}$$

where the quantity k given by [100, 101]

$$k = 0,0793 \frac{(M_1 + M_2)^{3/2}}{M_1^{3/2} M_2^{1/2}} \frac{Z_1^{2/3} Z_2^{1/2}}{\left(Z_1^{2/3} + Z_2^{2/3}\right)^{3/4}} \tag{4.37}$$

is often within the interval $0,1 < k < 0,2$. Merely in the special case of $Z_2 \gg Z_1$, with Z_1 comparable to 1, does k appreciably exceed $0,2$. For a representative value of k the *electronic stopping cross section cuts the nuclear cross section* near the energy E_c corresponding to $\varepsilon_c = 4,75$.

In the neighborhood of $v = v_1$ the electronic stopping has a maximum, upon which it decreases and gradually approaches the Bethe stopping formula. For small velocities of the particle ($\beta \ll 1$), the Bethe formula reduces to

$$-\left(\frac{dE}{dx}\right) = \frac{4\pi n Z_1^2 e^4}{m_e v^2} \cdot \ln\left(\frac{2m_e v^2}{I}\right) \tag{4.38}$$

where I is the mean excitation potential of the target ($I \cong Z_2 \cdot 10\,eV$) and n the electron density of the target given by

$$n = \frac{N_A \cdot Z_2 \cdot \rho}{A_2} \tag{4.39}$$

with the Avogadro number N_A and the mass number A.

4.6. SPUTTER YIELD

For a more detailed treatment of the electronic energy loss in the relativistic energy regime the reader is referred to [103].

4.6 Sputter yield

The first general predictive equation for the sputter yield Y of an incident ion with the energy E was given by Sigmund [6]

$$Y = \frac{0,042\, \alpha S_n(E)}{U_0} \qquad (4.40)$$

where α is a dimensionless factor that provides the proportion of energy from the incident ion back-reflected to be available for sputtering, $S_n(E)$ is the nuclear stopping cross section and U_0 is the surface binding energy per atom. Equation 4.40 is derived from the following three essential components:

- The ratio of energy deposition in the solid at the surface
- The fraction of energy back-reflected
- The rate of emission of atoms from the surface as a result of the back-reflected energy.

Since the ratio of energy deposition in the solid is directly and the emission rate of atoms from the surface is indirectly proportional to the atomic density, the *sputtering yield is independent on the atomic density*.

The basic equation of the sputtering yield has been developed by the years by additions and modifications of the relevant parameters. Three corrections have been made to equation 4.40:

- The addition of an electronic stopping contribution that grows in importance for higher energies and can be important for light primary ions
- A threshold effect that reduces the sputtering at low energies
- A target element-specific factor Q determined by fitting to experimental data

Thus, at normal incidence Y is given by [100, 101]

$$Y = \frac{0,042\, Q\alpha^*}{U_0} \frac{S_n(E)}{1 + AS_e(\varepsilon)} \left[1 - \left(\frac{E_{th}}{E} \right)^{1/2} \right]^s \qquad (4.41)$$

where U_0 is the surface binding energy in electron-volts, E_{th} is the threshold energy for sputtering, $S_e(\varepsilon)$ is the electronic stopping cross section in reduced energy units, A and s are fitting parameters. Calculated sputter yield values for Neon, Argon and Xenon ions can be found in [104].

Chapter 5

Experimental Setup

5.1 Ion generation and beam optics

The experimental setup for the IID measurements is shown in figure 5.1. A differential pumping system (cf. section 5.2) consisting of three *Turbo Molecular Pumps - TMP* (Group Pfeifer) and one *Ion Pump* provides a base pressure in the low 10^{-10} mbar range (without beam) inside of the experimental vessel. The total pressure in the system is measured with penning gauges and inside the UHV-chamber with two *Bayard-Alpert ionization gauges* (BA). The partial pressure in this chamber is measured with a *Residual Gas Analyzer - RGA* (BALZER QMS 112 with QMA 125 head). A gas injection system together with two needle valves is used to inject noble gases like He, Ne, Ar, Kr, Xe and other gases such as H_2, CH_4, C_2H_4, C_2H_6, CO, CO_2, N_2 for the purpose of calibration into the UHV-chamber (cf. section 5.3) and for the purpose of ion production into the ion gun. A dipole magnet is used for a mass-to-charge selection of the ions and by the help of three paired deflection plates the beam position can be changed.

Figure 5.2 shows a simulated ion beam passing through the experimental setup (cf. section 5.4.4).

5.1.1 Ion gun and first lens

5.1.1.1 Ionization probability

In the present case ions are produced from various gases by electron bombardment stimulated ionization. The ionization energy itself depends strongly on the attracting Coulomb force between the atomic core and the electron which should be removed [106].

Figure 5.3 shows the ionization energy for single ionization of the elements according to their atomic number.

CHAPTER 5. EXPERIMENTAL SETUP

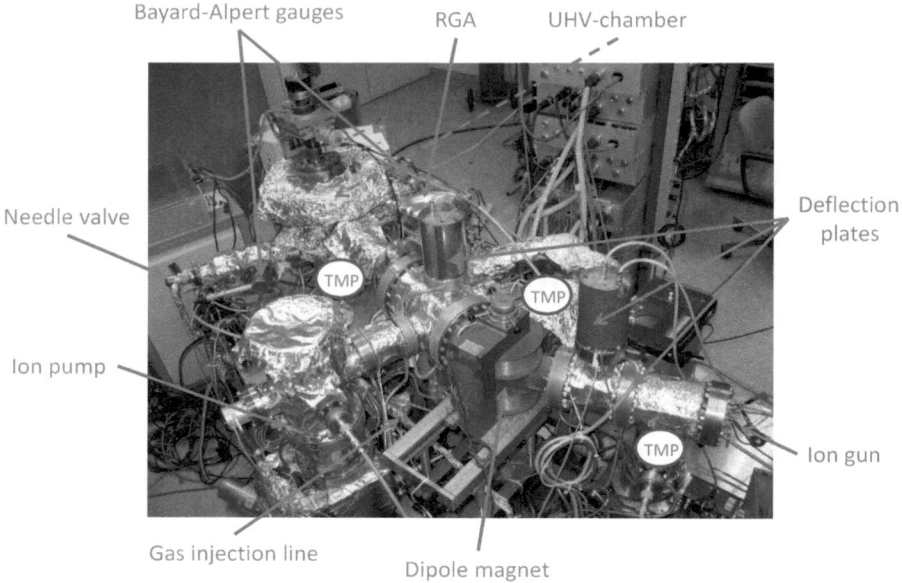

Figure 5.1: Schema of the experimental setup.

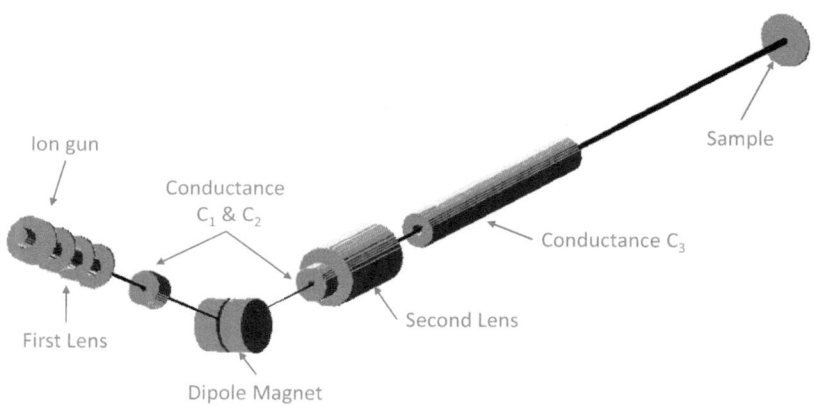

Figure 5.2: Ion optic simulation of the experimental system figured out with the SIMION program[105]: Ions created from various gases by electron bombardment are formed into an ion beam which has to pass a dipole magnet where the ions are selected after their mass-to-charge ratio. Hence only well defined ions can hit the sample inside of the UHV-chamber.

5.1. ION GENERATION AND BEAM OPTICS

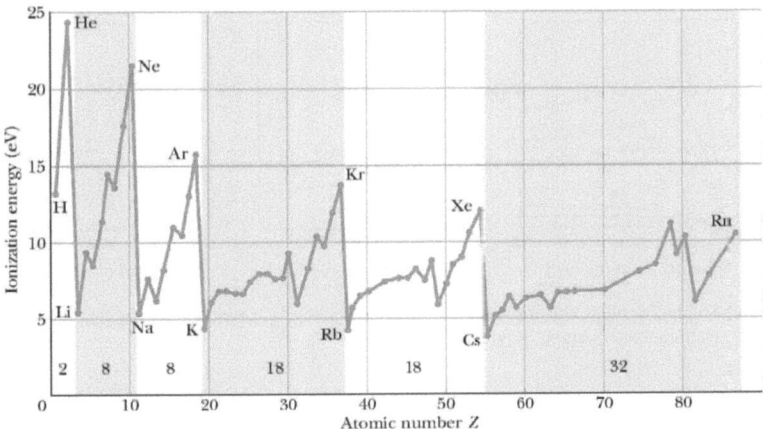

Figure 5.3: Single ionization energy of the elements according to their atomic number Z [106].

It can be seen that the ionization energy increases within a period of the periodic table (e.g. from H to He) due to the increasing atomic number Z (equal to the positive charge in the nucleus). Within a group (e.g. from He to Rn) in contrast the ionization energy decreases from top to bottom because the distance between the core and the electron becomes larger. The crossover from one period to the next (eg. from Helium to Lithium) results in a big decrease of ionization energy because the electron which should be removed is then located in a new shell and therefore bounded very slightly. For this reason noble gases have the highest ionization energy in each period.

The energy needed for double ionization (to remove a second electron from the same shell) increases compared to the one for single ionization because the atomic core is then doubly charged. Therefore the ionization cross section is smaller for higher grade ionization compared to single ionization.

The probability for single electron stimulated ionization is zero below a certain threshold energy[1], has a maximum between 3.3 and 5.7 times the threshold energy and decreases slowly after [107]. For the common gases the ionization curves present a broad maximum between 100 and 120eV electron kinetic energy [108]. Figure 5.4 shows the ionization energies for single-, double- and triple ionization on the basis of Helium, Neon, Argon, Krypton and Xenon.

---

[1]Threshold energies can be found in http://webbook.nist.gov/chemistry/form-ser.html

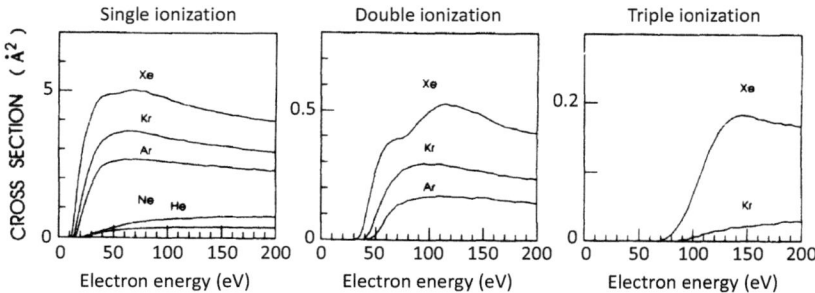

Figure 5.4: Single-, double- and triple ionization cross section as a function of the electron energy for Helium, Neon, Argon, Krypton and Xenon [107].

5.1.1.2 Spark discharge - Paschen curve

Paschen [109] found out in 1889 that the breakdown voltage V of two parallel plates in a gas as a function of the pressure p and gap distance d can be described by

$$V = \frac{a(pd)}{\ln(pd) + b} \tag{5.1}$$

where the constants a and b depend upon the composition of the gas. The graph of this equation is called "Paschen curve". By differentiating equation 5.1 with respect to (pd) and setting the derivative to zero, the minimum voltage can be found. This yields

$$pd = e^{1-b} \tag{5.2}$$

and predicts the occurrence of a *minimum breakdown voltage*. The composition of the gas determines both the minimum arc voltage and the distance at which it occurs.

Figure 5.5 shows Paschen curves for some gases which are used in the ion gun. It should be noted that Neon has the lowest strike-over voltage.

5.1. ION GENERATION AND BEAM OPTICS

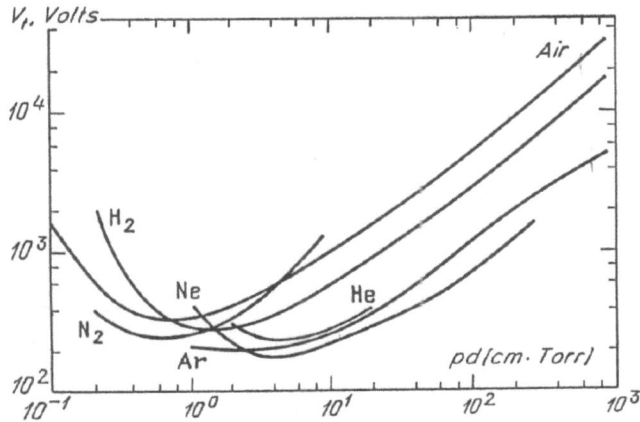

Figure 5.5: Paschen curves for various gases [110].

5.1.1.3 Assembling and function

A schematic of the ion gun is shown in figure 5.6: Electrons are emitted from a heated wolfram filament (5, 6) and accelerated towards the grid (10). On their way they collide with the gas, injected through the gas injection system (9), and produce ions of different charge states according to the ionization probability.

After ionization the ions with the potential energy $E = q \cdot U$, given by the high voltage on the grid, have to be extracted by the help of an extraction voltage (7).

The optimum extraction voltage V_E varies with the accelerating voltage V_A (given by the potential difference between grid and ground) and also varies slightly with emission current and gas pressure in the ion gun. However, the optimum voltage for extraction is approximately a constant proportion of the accelerating voltage and is given by

$$V_E \cong 0,95 \cdot V_A \tag{5.3}$$

After their extraction the ions are accelerated to ground potential and have to pass an Einzel lens (8) (two grounded cylinders interrupted by a cylinder on high voltage) which focuses the beam into the center of a dipole magnet for a mass-to-charge selection of the ions. The high voltage V_{FL} which has to be applied on the first lens (FL) is also approximately a constant proportion of the accelerating voltage and is given by

$$V_{FL} \cong 0,56 \cdot V_A \tag{5.4}$$

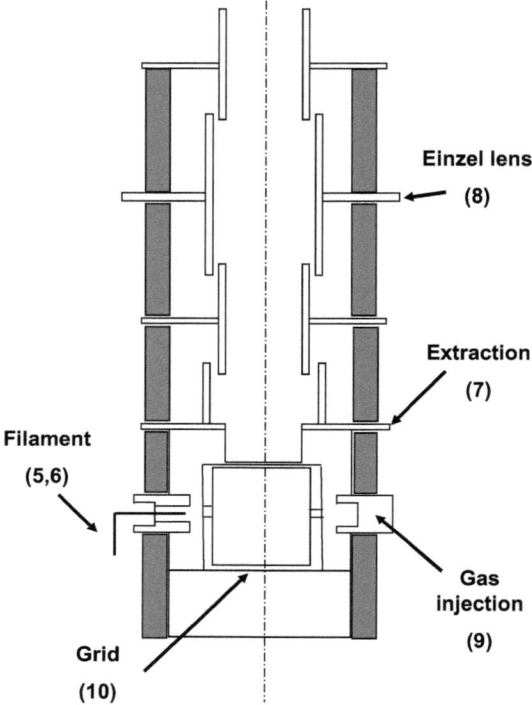

Figure 5.6: Schema of the ion gun.

Figure 5.7 shows a simulation (cf. section 5.4.4) of the potential in the ion gun and the first lens for a given grid voltage of 5kV, an extraction voltage of 4,75kV and a focusing voltage for the lens of 2,8kV for the case of Ar^+-ions.

5.1.2 Dipole magnet

In the dipole magnet the so called Lorentz force F_L

$$\vec{F}_L = q \cdot \vec{v} \times \vec{B} \tag{5.5}$$

and the centrifugal force F_C

$$F_C = \frac{mv^2}{r} \tag{5.6}$$

which both act on a moving particle (with the speed $\vec{v}$, mass m and charge q), have to be in equilibrium.

5.1. ION GENERATION AND BEAM OPTICS

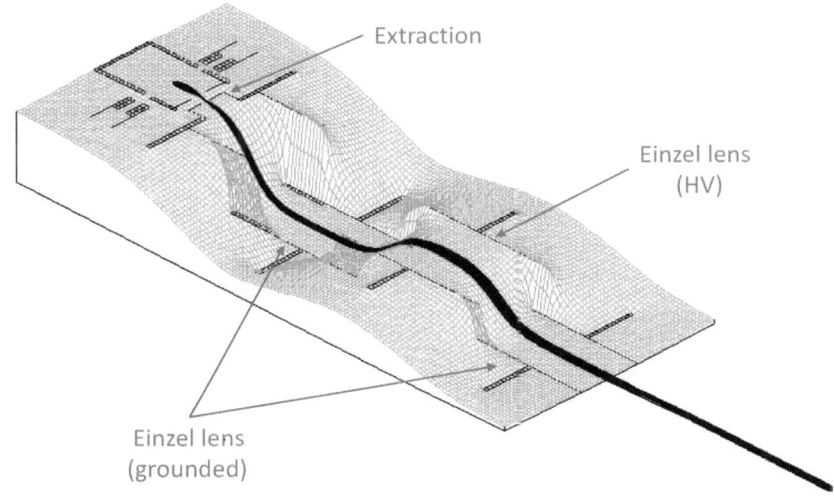

Figure 5.7: Potential energy view in the ion gun for the case of Ar^+-ions simulated with SIMION.

Considering that $\vec{v}$ is perpendicular to the magnetic field $\vec{B}$, F_L can be also written as

$$F_L = q \cdot v \cdot B \tag{5.7}$$

Also considering that potential energy E_{pot} is directly converted into kinetic energy E_{kin}

$$E_{pot} = q \cdot U \tag{5.8}$$
$$E_{kin} = \frac{mv^2}{2} \tag{5.9}$$

(U is the acceleration voltage of the particle), the equation for the necessary B-field for the bending of the particle track in the dipole magnet is given by

$$B = \frac{1}{r} \cdot \sqrt{\frac{2Um}{q}} \tag{5.10}$$

Therefore the *B-field selects particles by their mass-to-charge ratio*.

5.1.2.1 Dipole geometry

The geometry of the magnet is fixed by the angel α ($\alpha = 15, 13°$) between the line A-B and the line between the beam pipe (C_2) and the center of the pole shoe. As it is shown in figure 5.8, the nominal beam trajectory trough the two pipes for a particle which has been shot into the center of the dipole requires a circular trajectory with the bending radius R.

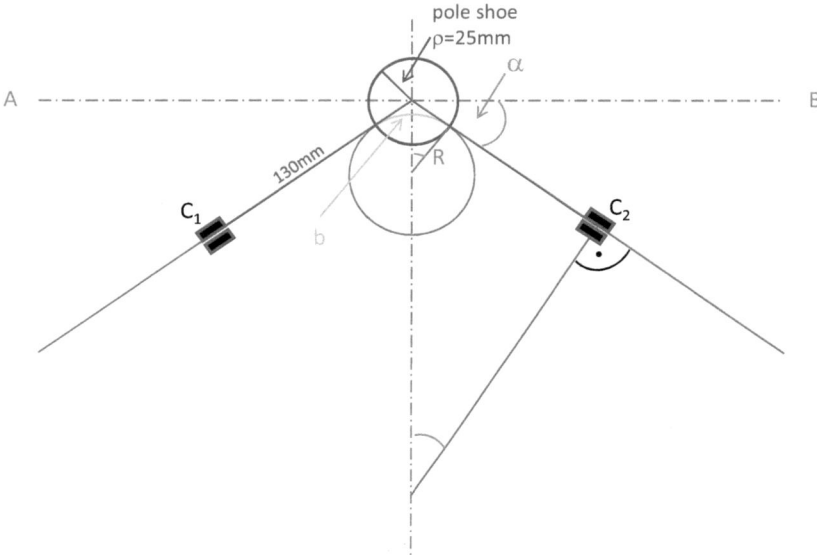

Figure 5.8: The geometry of the dipole magnet is fixed by the angel α.

According to equation 5.10 this would state for the B-field

$$B_{required} = \frac{1}{R} \cdot \sqrt{\frac{2Um}{q}} \qquad (5.11)$$

where R is related to ρ (the radius of the pole shoe) and α by

$$R = \rho \cdot \tan(90 - \alpha) \qquad (5.12)$$

5.1.2.2 Focusing effect

If a particle has not been shot exactly into the center of the magnet, the deflection angle $(2 \cdot \gamma)$ depends on the point of impact (ρ, φ) on the dipole (cf. figure 5.9).
γ is related to ρ and φ by

$$\tan \gamma = \frac{\rho \cdot \cos \varphi}{R + \rho \cdot \sin \gamma} \qquad (5.13)$$

Hence additionally the dipole has a focusing and a de-focusing effect, depending on the place of impact as it is shown in figure 5.10.

5.1. ION GENERATION AND BEAM OPTICS

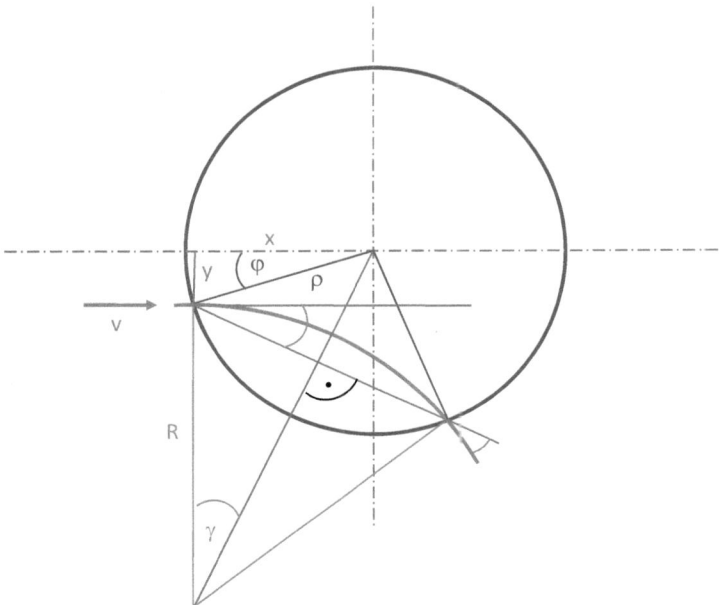

Figure 5.9: Deflection angle of the dipole magnet for a particle which has not been shot into the center of the dipole.

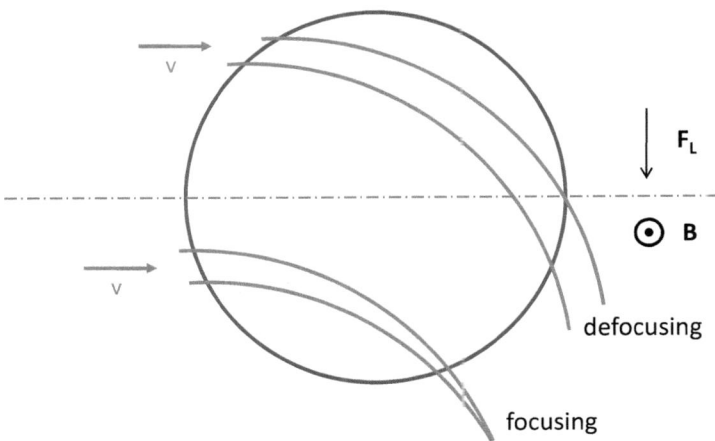

Figure 5.10: Focusing and de-focusing effect of a dipole magnet: For a given B-field normal to the plane a beam with a certain elongation can be focused and de-focused depending on the place of impact.

5.1.3 Second lens

After the dipole magnet, the beam has to pass a second electrostatic lens (SL) to be focused on the Faraday cup (cf. section 5.4.3). The optimal voltage setting for this lens is also approximately a constant proportion of the accelerating voltage and is given by

$$V_{SL} \cong 0,45 \cdot V_A \quad (5.14)$$

Figure 5.11 shows the voltage settings for the extraction and for the two lenses as well as the current setting for the dipole magnet in case of Ne$^+$-ions. Note that the magnet current slightly differs from the square root behavior (dashed line) of equation 5.10.

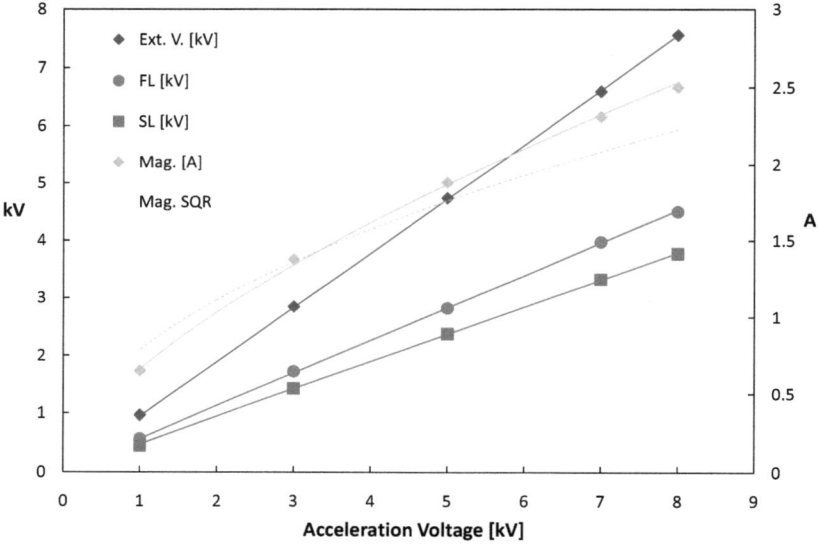

Figure 5.11: Ion gun settings for Ne$^+$-ions.

5.1.4 Beam monitoring and deflection

Two beam monitors (four isolated, circular arranged stainless steel segments) at the entrance and exit aperture of the magnet measure the beam position. For a well aligned beam the four segments should measure approximately the same ion current. In order to make corrections, the beam can be deflected by the help of paired deflection plates. One pair is mounted right after the first lens in order to align the ion beam with the entrance aperture in front of the dipole magnet. An other pair is mounted at the exit aperture of the dipole magnet and a last one is mounted right after the second lens. These are used for small corrections of the beam position on the sample. The vertical deflection plates are

10mm and the horizontal plates 20mm apart from each other. The potential difference between one pair can be set by a potentiometer between 0V and 1000V.

5.2 Differential pumping system

Inside of the experimental vessel a very low pressure in the 10^{-10}mbar range is required to prevent the sample surface from changes due to reactions with the residual gas. Since inside the ion gun the pressure has to be in the 10^{-6}mbar range to obtain a reasonable ion current, a differential pumping system becomes necessary.

For an injected N_2 quantity of $1,6 \times 10^{-4}$mbar·l/s the measured pressure inside the ion gun is approx. 1×10^{-6}mbar, inside the magnet approx. $1,3 \times 10^{-8}$mbar, inside the lens chamber approx. 3×10^{-9}mbar and inside the UHV-chamber approx. 2×10^{-10}mbar (cf. figure 5.12).

The measured pressure rise ΔP_4 inside the UHV-chamber can be calculated as a function of the injected gas quantity Q into the ion gun:

The TMP with the pumping speed S_1 of approximately 160l/s (N_2 equiv.) is pumping the ion gun vacuum chamber which is connected by means of the conductance C_1 ($l = 2$cm, $D = 0,6$cm $\rightarrow$ $C_{1;N_2} \approx 1,31$l/s) with the magnet vacuum chamber.

The flow rate C of such a conductance (long pipe, molecular flow with P< 10^{-3}Torr) is given by [111]

$$C_{(Gas\ 20°C)}[\frac{l}{s}] = 12,1 \cdot \frac{D^3[cm]}{L[cm]} \cdot \sqrt{\frac{28}{m}} \qquad (5.15)$$

with D and L the diameter and the length of the pipe, respectively.

This chamber is pumped by the TMP with S_2 ($\cong S_1$) and connected by means of another conductance C_2 ($\cong C_1$) to the second lens vacuum chamber right after the magnet. Here the pressure is already in the 10^{-9}mbar range and therefore an *ion pump* is pumping this chamber with the pumping speed S_3 of approximately 180l/s (N_2 equiv.). This chamber is connected by means of the conductance C_3 ($l = 11,5$cm, $D = 1$cm $\rightarrow C_{3;N_2} \approx 1,05$l/s) with the experimental UHV-chamber which is pumped by the calibrated TMP with S_4 (approximately 32l/s (N_2 equiv., cf. section 5.3.1).

From the following system of equations (knowing the values of $Q, S_1, S_2, S_3, S_4, C_1, C_2, C_3$)

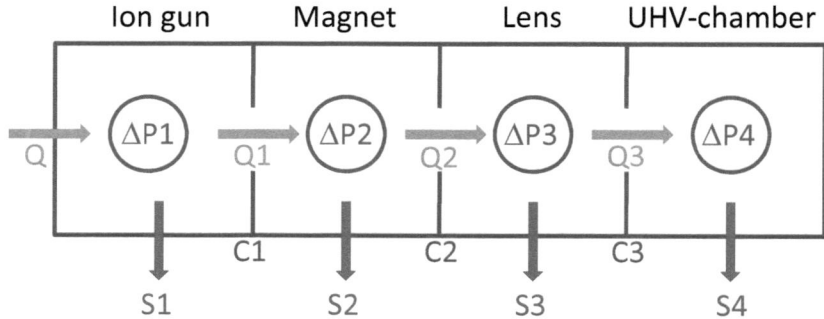

Figure 5.12: Schema of the differential pumping system. The pressure rise ΔP_4 in the UHV-chamber can be calculated as a function of the injected gas quantity Q into the ion gun.

$$Q = \Delta P_1 S_1 + \underbrace{(\Delta P_1 - \Delta P_2) \cdot C_1}_{Q_1} \quad (5.16)$$

$$(\Delta P_1 - \Delta P_2) \cdot C_1 = \Delta P_2 S_2 + \underbrace{(\Delta P_2 - \Delta P_3) \cdot C_2}_{Q_2} \quad (5.17)$$

$$(\Delta P_2 - \Delta P_3) \cdot C_2 = \Delta P_3 S_3 + \underbrace{(\Delta P_3 - \Delta P_4) \cdot C_3}_{Q_3} \quad (5.18)$$

$$(\Delta P_3 - \Delta P_4) \cdot C_3 = \Delta P_4 S_4 \quad (5.19)$$

ΔP_4 is given by:

$$\begin{aligned}\Delta P_4 = Q \cdot \{ &S_1 + S_2 + S_3 + S_4 + \frac{S_1}{C_1} \cdot (S_2 + S_3 + S_4) + \frac{S_4}{C_3} \cdot (S_1 + S_2 + S_3) + \\ &+ \frac{S_1 S_2}{C_1 C_2} \cdot (S_3 + S_4) + \frac{S_1 S_4}{C_1 C_3} \cdot (S_2 + S_3) + \frac{S_3 S_4}{C_2 C_3} \cdot (S_1 + S_2) + \\ &+ \frac{S_1}{C_2} \cdot (S_3 + S_4) + \frac{S_2}{C_2} \cdot (S_3 + S_4) + \frac{S_1 S_2 S_3 S_4}{C_1 C_2 C_3} \}^{-1}\end{aligned} \quad (5.20)$$

5.3 System calibration

5.3.1 Calibration of the pumping speed

The pumping speed S_{sys} of a TMP and the sensitivity factors ς_{RGA} of a RGA are specific for various gases and differ for several devices. A pre-calibration of these components is necessary. The calibration of the pumping speed is done through a pipe with a well known conductance (8.4 l/s N_2

5.3. SYSTEM CALIBRATION

equiv.) and two BA-gauges (cf. figure 5.13). During the calibration the connection valve to the beam system is closed.

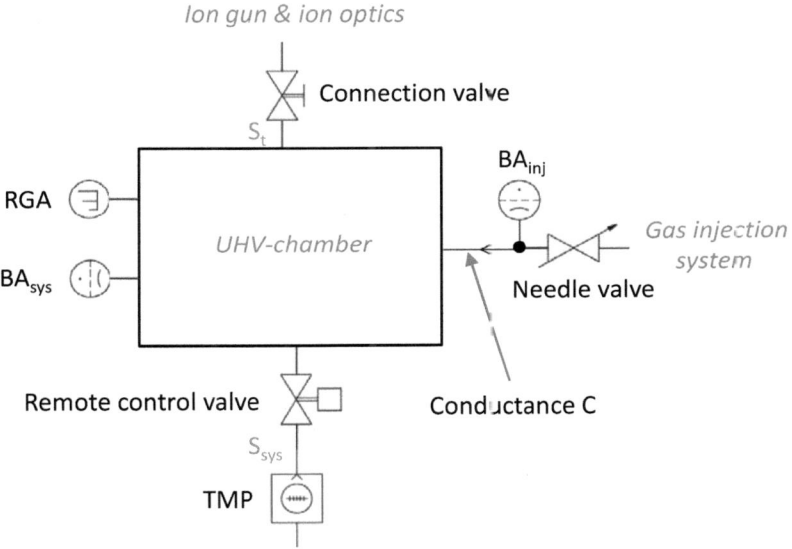

Figure 5.13: Schematic of the UHV-chamber.

During injection of a certain amount of gas into the UHV-chamber a pressure rise ΔP_{inj} at the BA_{inj}-gauge before the conductance and a pressure rise ΔP_{sys} at the BA_{sys}-gauge after the conductance inside the UHV-chamber is measured. The pumping speed can then be determined by:

$$\Delta P_{sys} \cdot S_{sys} = C(\Delta P_{inj} - \Delta P_{sys}) \tag{5.21}$$

hence

$$S_{sys} = C \cdot \left(\frac{\Delta P_{inj}}{\Delta P_{sys}} - 1\right) \tag{5.22}$$

If the valve to the system pump is closed ($S_{sys} = 0$), the two gauges should measure the same pressure. Since they differ due to their different intern sensitivity factors ς_i and ς_s, a correction factor $k = \frac{I_{s0}}{I_{i0}}$ is introduced and measured for each gas. For $S_{sys} = 0$ this applies:

$$\Delta P_{inj} = \Delta P_{sys} \tag{5.23}$$

$$\frac{I_{i0}}{\varsigma_i \cdot i_e} = \frac{I_{s0}}{\varsigma_s \cdot i_e} \tag{5.24}$$

$$\frac{\varsigma_s}{\varsigma_i} = \frac{I_{s0}}{I_{i0}} \tag{5.25}$$

where I_{i0} and I_{s0} are the ion currents on the BA-gauges, measured at $S_{sys} = 0$ and i_e is the emission current (set to 1mA for each gauge). The relevant pumping speed of the TMP in the UHV-chamber can then be calculated for each gas using equation 5.26.

$$S_{sys} = C \cdot (\frac{I_i}{I_s} \cdot k - 1) \tag{5.26}$$

I_i and I_s are the ion currents on the BA-gauges during gas injection. Hence S_{sys} is determined independent of the gauge sensitivities.
Figure 5.14 shows the measured ion currents on the two BA-gauges, the resulting k-factor and its average.

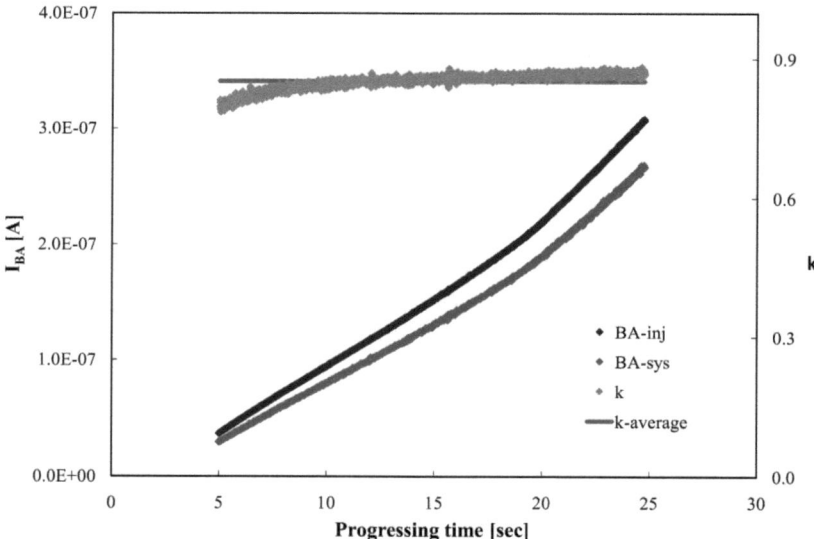

Figure 5.14: Measured ion currents on the two BA-gauges, the resulting k-factor and its average.

According to equation 5.26 the calibration of the pumping speed for a certain gas has been done

5.3. SYSTEM CALIBRATION

by a gradually injection of the gas into the UHV-chamber starting from a system base pressure of 1×10^{-10} mbar up to 1×10^{-6} mbar. Figure 5.15 shows the calculated pumping speed in case of argon injection. The decrease of pumping speed in the high pressure range is due to the fact that starting from this pressure the BA-gauges do not respond anymore linearly to a further pressure increase.

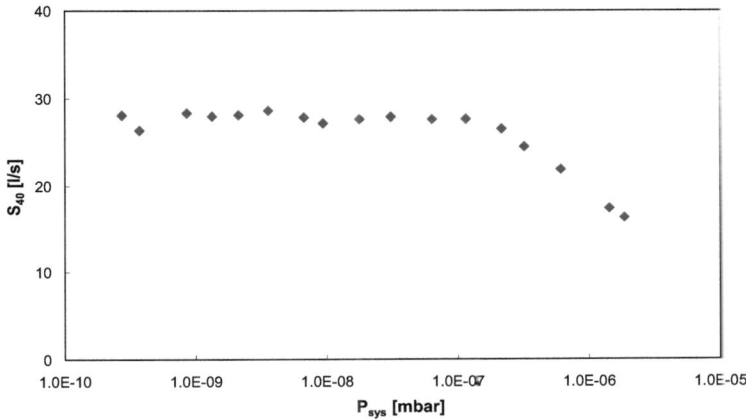

Figure 5.15: Calculated pumping speed in the case of argon injection.

Since the pumping speed is proportional to the flow rate through the conductance (cf. equation 5.22), a fit can be done with the measured data points as it is shown in figure 5.16. It can be seen that the measured values correspond very good with the fit with the exception of light gases, e.g. hydrogen and helium. This is due to the fact that the compression factor of the TMP is not that efficient for these light gases.

For an estimation of the pumping speed S_t through the connection tube, the valve between the two systems has been opened and the valve to the system pump has been closed. In case of helium injection into the UHV-chamber a value of 3,5 l/sec for S_t has been obtained. This value accounts only for about 4% of the total pumping speed and was therefore neglected.

5.3.2 RGA calibration

Both, the BA-gauge and the RGA are measuring the current of collected ions. On one hand the current is proportional to the total pressure and on the other hand it is proportional to the partial pressure in the system.

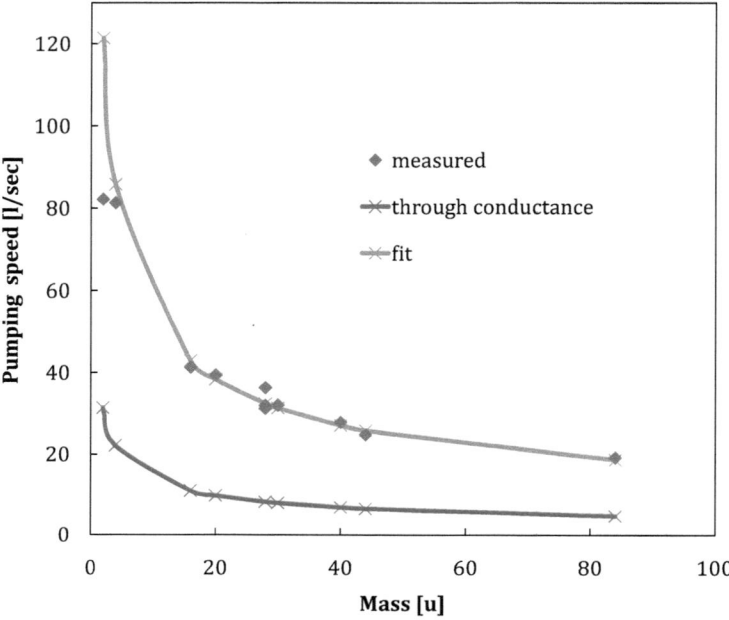

Figure 5.16: Measured and fitted values of the pumping speed in the case of argon injection. The flow rate C "through conductance" was derived for different masses m from $C = 8,4 \cdot \sqrt{28/m}$. The measured pumping speeds were fitted to this equation by a multiplication factor which value is 3,86.

$$I_{BA,RGA} = P_{BA,RGA} \cdot \varsigma_{BA,RGA} \cdot i_e \qquad (5.27)$$

where i_e is the emission current of these two devices (set to 1mA in both cases).

During the injection of a single gas, the total pressure at the BA-gauge and at the RGA is the same. Hence it is possible to calibrate the sensitivity ς_{RGA} of the RGA for various gases (assumed that the BA gauge is pre-calibrated). Plotting the current measured at the RGA versus the base pressure in the system shows a function which slope yields the product of sensitivity factor times emission current (given in A/mbar) as it is shown in figure 5.17. Hence $\varsigma_{RGA} \cdot i_e$ is given by

$$\varsigma_{RGA} \cdot i_e = \frac{I_{RGA}}{P_{BA}} \qquad (5.28)$$

5.3. SYSTEM CALIBRATION

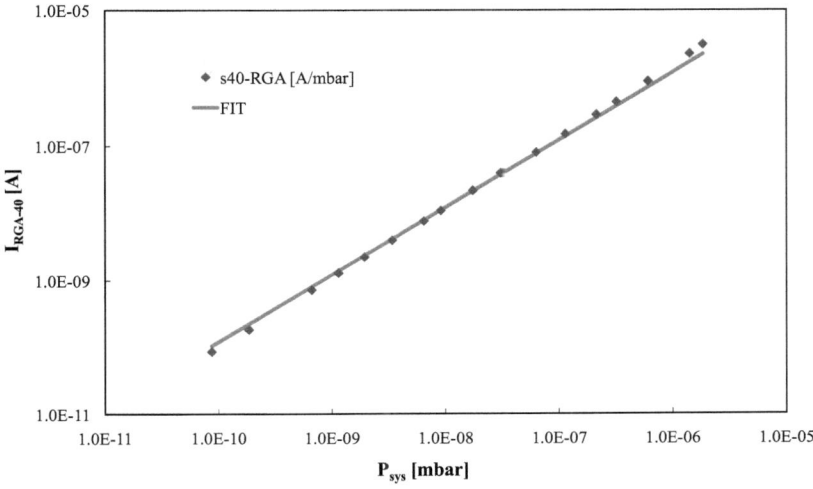

Figure 5.17: RGA sensitivity during argon injection. The slope of the linear fit through the data points represents $\varsigma_{RGA} \cdot i_e$ in units of A/mbar.

5.3.2.1 Sensitivity factors and evaluation of ion currents

To evaluate the desorption contributions of H_2, CH_4, CO, C_2H_4, C_2H_6 and CO_2 it is necessary to calibrate the RGA also for the cracking patterns of these gases. The results of this calibration are shown in table 5.1.

The specific ion current contributions of each gas can be evaluated using the following system of equations:

$$
\begin{aligned}
I_{H_2} &= I_2 \\
I_{CH_4} &= I_{15} \\
I_{C_2H_6} &= I_{30} \\
I_{CO_2} &= I_{44} \\
I_{C_2H_4} &= (I_{27} - I_{C_2H_6}/0,671)/0,666 \\
I_{CO} &= I_{28} - I_{C_2H_4} - I_{C_2H_6}/0,222 - I_{CO_2} \cdot 0,409
\end{aligned}
$$

Together with equation 4.8 and the proper sensitivity factors the desorption yields for the mentioned gases can be calculated.

Gas	Mass	$\varsigma_{RGA} \cdot i_e$ [A/mbar]	Contribution to main peak
H_2	2	3,6	100%
He	4	1,8	100%
CH_4	16	1,9	100%
CH_4	15	1,6	84,4%
Ne	20	5,9	100%
N_2	28	1,9	100%
N_2	14	0,3	13,6%
CO	28	1,9	100%
CO	12	0,2	7,8%
C_2H_4	28	1,5	100%
C_2H_4	27	1,1	66,6%
C_2H_6	30	0,6	22,2%
C_2H_6	28	2,5	100%
C_2H_6	27	0,8	33,1%
Ar	40	1,1	100%
CO_2	44	1,2	100%
CO_2	28	0,5	40,9%
Kr	84	0,9	100%

Table 5.1: RGA sensitivities for injected gases and their cracking patterns.

5.4 Modifications and additions to the setup

5.4.1 Design of a new ion gun power supply

The ion gun produced by Edwards was operated in the past with a modified power supply of a BA-gauge [108]. Since the documentation of this power supply was not updated and the filament lifetime was too short to obtain reproducible results, a new power supply had to be designed.

Although the geometric assembly of the ion gun differs very much from a BA-gauge, e.g. the gauge filament with 0,18mm in diameter is much longer than the filament of the ion gun (a thoriated tungsten W99/Th1 wire with 0,15mm in diameter), the electronic characteristics of both devices have been recorded and compared.

For these measurements both, the ion gun and the gauge were not floated with high voltage. Electrons were accelerated from the filament (+50V) towards the grid which was biased at +200V above ground level (cf. figure 5.20) and in both cases the electron repeller was set to 0V. Figure 5.18 shows the electron emission current for the gauge and the ion gun under these conditions.

Measurements have shown that for the same potential difference (150V) between filament and grid the ion gun shows a higher emission current, if the repeller and the filament were kept on the same potential as it is shown in figure 5.19.

5.4. MODIFICATIONS AND ADDITIONS TO THE SETUP

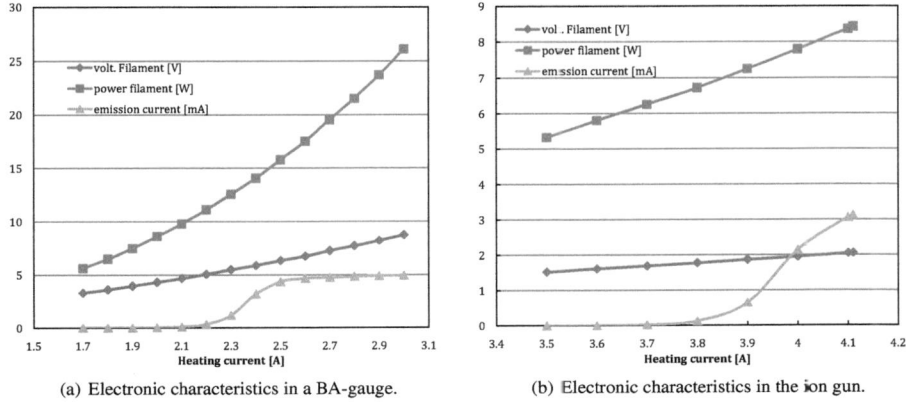

(a) Electronic characteristics in a BA-gauge.

(b) Electronic characteristics in the ion gun.

Figure 5.18: Comparison of BA-gauge and ion gun (repeller: 0V; filament: +50V; grid: +200V).

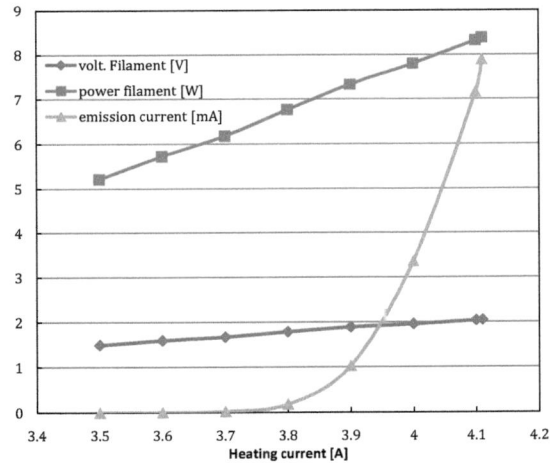

Figure 5.19: Electronic characteristics in the ion gun without repeller (repeller: +50V; filament: +50V; grid: +200V).

The requirements for the new power supply are shown in figure 5.20. The high voltage applied on the grid serves as a reference for all other units and defines the kinetic energy of the ions. A detailed scheme of the new electronic board[2] is given in chapter A.5.

---
[2]designed by David Porret - former AT/VAC/ICM section at CERN

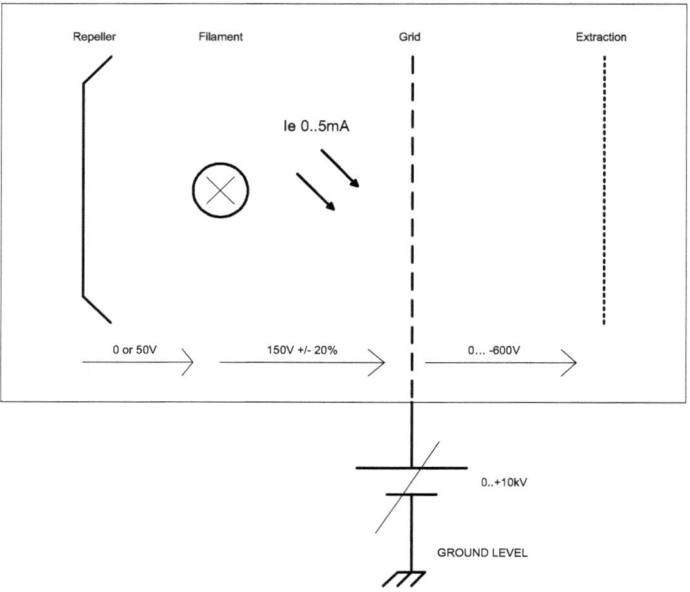

Figure 5.20: Requirements for the ion gun power supply.

5.4.2 Pumping speed reduction in the UHV-chamber

In section 4.4 it was shown that η_{min} is direct proportional to the pumping speed and a decrease of the pumping speed ensures the measurement of smaller desorption yields.

On the other hand, the rise time of the signal is indirect proportional to the pumping speed (cf. equation 4.6). Hence a higher ion dose due to a longer bombardment of the sample is required to get a steady state desorption signal.

For these reasons the pumping speed in the UHV-chamber has been reduced by a factor of 5 by means of a fixed conductance. The diameter of the conductance is calculated in the following way:

$$S_n = \frac{S_a}{5} \tag{5.29}$$

and

$$\frac{1}{S_n} = \frac{1}{S_a} + \frac{1}{C} \tag{5.30}$$

5.4. MODIFICATIONS AND ADDITIONS TO THE SETUP

Where S_a is the actual pumping speed, S_n the new required pumping speed and C the flow rate through the orifice, hence

$$C = \frac{S_a}{4} \tag{5.31}$$

Since the flow rate C for gases (mass m) through orifices under molecular flow (below 10^{-3}Torr) is given by [111]

$$C_{(Gas\,20°C)}[\frac{l}{s}] = 11,6 \cdot \frac{\pi D^2[cm]}{4} \cdot \sqrt{\frac{28}{m}} \tag{5.32}$$

the required diameter D (≈ 1.66cm, N_2 equiv.) is given by

$$D = \sqrt{\frac{S_a \cdot \sqrt{m}}{11,6 \cdot \pi \cdot \sqrt{28}}} \tag{5.33}$$

Due to this change in S_{sys} the minimum achievable desorption coefficients are calculated from equation 4.19 for the various gases and shown in table 5.2 ($I_{min} = 10^{-11}$A and $I_{max} = 10^{-7}$A).

Element	Mass	S [l/s]	$S_{RGA} \cdot i_e$ [A/mbar]	η_{min}
H_2	2	82,2	3,6	9,0E-03
He	4	81,3	1,8	1,8E-02
CH_4	16	42,9	1,9	8,7E-03
Ne	20	38,4	5,9	2,6E-03
N_2	28	32,4	1,9	6,9E-03
CO	28	32,4	1,9	6,9E-03
C_2H_4	28	32,4	1,7	7,7E-03
C_2H_6	30	31,3	2,5	5,0E-03
Ar	40	27,1	1,1	1,0E-02
CO_2	44	25,9	1,2	8,5E-03
Kr	84	18,7	0,9	8,6E-03

Table 5.2: Minimum achievable desorption coefficients calculated for various gases.

5.4.3 Faraday cup for beam positioning

It became necessary to monitor the beam inside the UHV-chamber in order to know the exact place of beam impact on the sample. Therefore a Faraday cup (figure 5.21) has been constructed and mounted on a rotatable sample holder together with three samples.

The cup consists of a copper tube which is mounted electrically isolated by means of two ceramic washers on a 45mm × 45mm stainless steel plate. Ions which enter through an aperture of 600μm in diameter are collected inside the copper tube. The defined position of the aperture hole (15mm by 15mm from the lower left corner of the stainless steel plate) is used to adjust the beam with normal incidence.

The sample can be scanned with the beam by moving the sample holder in vertical and horizontal direction. For a beam cross section of 50mm² and a uniform ion density the ratio between the current measured in the Faraday cup and the current measured on the sample, called as R_{FC}, has to be approximately 0,0056. Due to the geometry of the filament-extraction system (different distances between filament and grid due to the bending of the filament), the ion density of the beam is not uniform and shows a maximum in the middle of its cross section.

It figured out that this ratio is a good tool for the control of the ion density of the beam over its cross section. The ion density of the beam can be changed by the help of the second lens (SL): If R_{FC} is set within 0,003 and 0,004 the beam has a very broad maximum, on the contrary if R_{FC} is set within 0,008 and 0,009 the beam has a sharp maximum of ion density in the middle of its cross section.

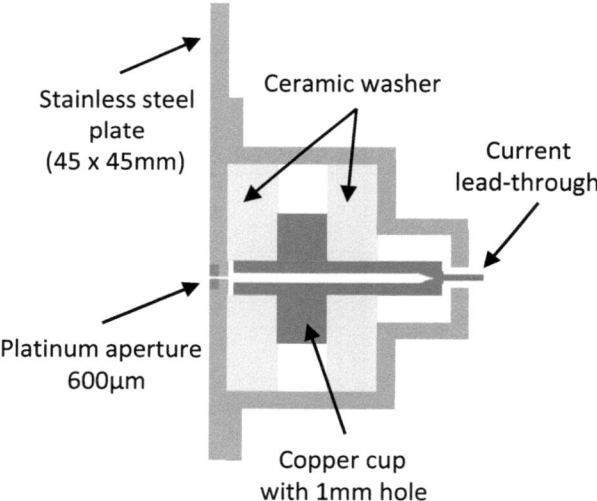

Figure 5.21: Cross section of the Faraday cup.

5.4.4 Ion optic simulations with the SIMION program

SIMION (version 7.0) was used to estimate the values for the extraction and lens voltages in equations 5.3, 5.4 and 5.14 as well as for the required magnetic field of the dipole magnet given by equation 5.11. SIMION defines magnetic potentials in *Mags* (Gauss times grid units) [105]. The required B-field in units of Mags can be calculated using the following equation:

$$B[Mags] \approx 15,57 \cdot \sqrt{m \cdot U} \cdot \frac{d}{100} \qquad (5.34)$$

where m is the weight of the atom (in atomic units) and U the acceleration voltage.

For the simulation it was assumed that the ions right after their creation were just having thermal kinetic energy ($E_{therm} = \frac{3}{2} \cdot k \cdot T$) of about 38meV at 300K.

5.5 Measurement procedure

After their cleaning treatment, e.g. degreasing followed by alkaline etching and rinsing in demineralized water (for a detailed description see chapter A.1), the samples were enveloped in aluminium paper and brought directly to the experimental setup.

Three samples together with the Faraday cup are mounted on a 360° rotatable sample holder under atmospheric pressure. After an initial pump down to 10^{-7}mbar a bakeout for 24 hours at 250°C was started followed by a degassing of the gauges and the RGA. An ultimate pressure in the low 10^{-10}mbar range inside of the UHV-chamber was achieved.

Figure 5.22 shows the beam control devices and the numbers which refer to the mentioned numbers in the following paragraphs.
Before operation the voltage between filament and grid was fixed to 120V with a potentiometer on the electronic board of the ion gun power supply (cf. figure A.5) to ensure the maximal ionization cross section.

The gas pressure inside the ion gun is measured with a Penning gauge and displayed at (1). During operation this pressure is set to approximately 1×10^{-6}mbar for all injected gases by the help of a needle valve which separates the gas injection system from the ion gun.
Then the grid voltage (2) is set to a value which defines the kinetic energy of the ions. According to equations 5.3, 5.4 and 5.14 the voltages of the extraction (3), first (4) and second lens (5) are fixed on the related power supplies.
Then the emission current on the ion gun filament is increased with the potentiometer (6) to a value of 2,5mA (maximum 5mA) and the current on the magnet power supply is tuned with the potentiometer (7) to obtain a maximum ion current in the Faraday cup. This current can be further increased by optimizing the beam position before and after the magnet, displayed on the beam position monitors (8), by changing the applied voltage on the deflection plates (9).

For each energy the beam has to be readjusted on the Faraday cup and the ratio R_{FC}, mentioned in chapter 5.4.3, has to be always the same. After the adjustment, the beam is totaly deflected by the help of the first deflection plates and the sample is rotated into beam position. The impact position on the sample itself can be changed in horizontal and vertical direction by the help of two millimeter

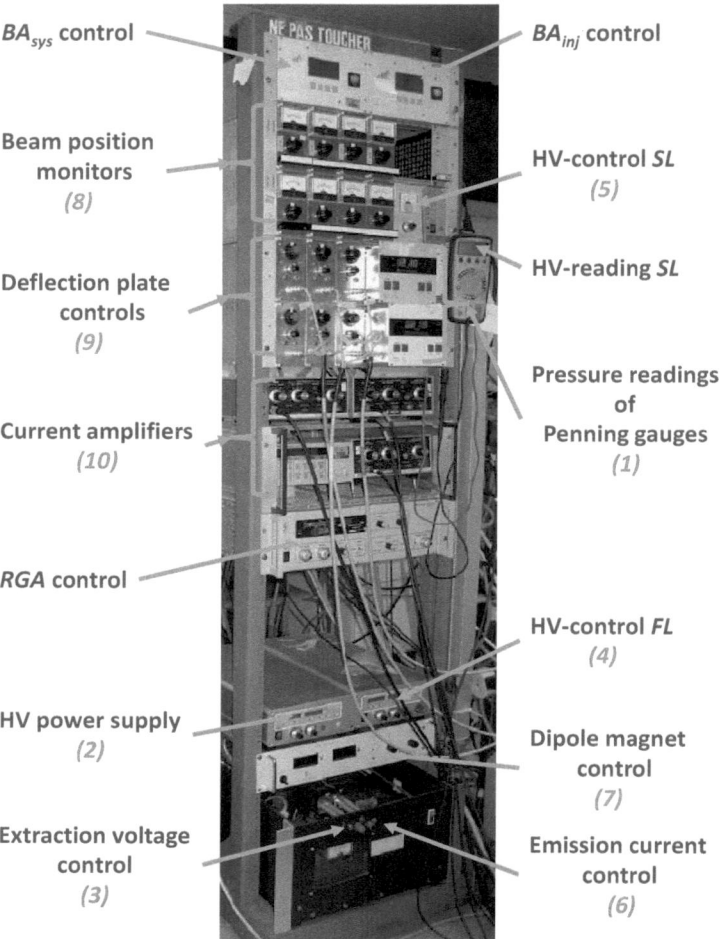

Figure 5.22: Beam control rack (HV = High Voltage).

screws.

During irradiation the pressure increases (measured as small ion currents on the gauges) due to ion bombardment are amplified (10) and recorded. To avoid the emission of secondary electrons during bombardment, the sample is biased at +18V.

5.5. MEASUREMENT PROCEDURE

5.5.1 Energy and mass dependent measurements

The dose dependance of the ion-induced desorption yield is a strong limitation for these measurements. At an accumulated ion dose between 10^{14} and 10^{15} ions/cm^2, the desorption yield starts to decrease due to an incipient cleaning of the sample as it is shown in figure 5.23.

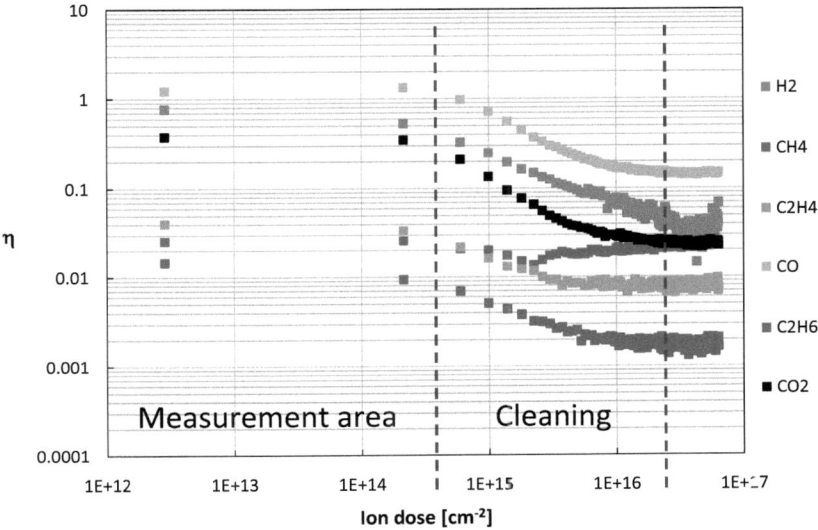

Figure 5.23: Dose dependance of the desorption yield during ion bombardment.

Therefore the maximum applied ion dose has to be below or equal to this value during the energy dependent measurements of the desorption yield. This is achieved by:

- Using short time pulses with a duration between 5 and 15 seconds, depending on the rise time of the signal (which is indirect proportional to the pumping speed, cf. equation 4.6).

- Using small beam currents of about 2×10^{-8}A [3].

- Setting the ratio R_{FC}, mentioned in section 5.4.3, to a value within 0,003 and 0,004 to guarantee a broad maximum in ion density over the beam cross section.

In order to ensure always to be at the peak maximum, the RGA was tuned manually to the masses 2, 15, 27, 28, 30 and 44. For each mass a short ion pulse desorption signal was recorded on the RGA (green) together with the ion current measured on the sample (red) and the total pressure measured with the BA-gauge (blue) as it is shown in figure 5.24.

---
[3] Since the ion current was very small, also the desorption signals were very small. Thus the current amplifier of the RGA was operated in the "zero suppression mode" to make this small signals visible.

CHAPTER 5. EXPERIMENTAL SETUP

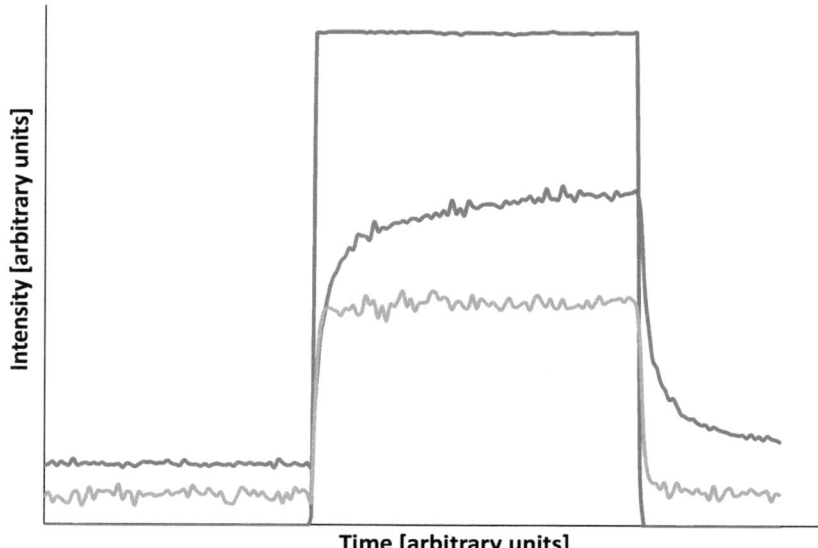

Figure 5.24: Short ion pulse desorption signals.

The samples were irradiated with all different ions starting from 1keV up to 7keV (in 2keV steps). For heavy ions, e.g. Kr^+ and Xe^+, it was not possible to exceed 5keV due an overheating of the magnet coils.

The frequent change of samples due a lack of non irradiated sample area made it impossible to compare desorption yields for different ions. Therefore the yields were re-measured in one run for each ion species at 3keV on one sample. These desorption yields were used as a normalization factor for the results obtained before. In order to get better signals, the ion current was risen for these measurements to approximately 8×10^{-8}A.

5.5.2 Dose dependent measurements

The measurements of the dose dependance of the desorption yield was done in a different way. Here a very intense ion beam with a sharp maximum in its ion density was used. Thus the ratio R_{FC}, mentioned in section 5.4.3, was set to a value within 0,008 and 0,009.

In order to have an almost simultaneous recording of the different mass signals the RGA was remotely controlled by a dedicated LabView programm which was adapted from an existing version of a similar experimental setup. The time between two recorded spectra was set to 20 seconds.

56

Chapter 6

Results and Discussion

6.1 Results

6.1.1 Influence of the ion nature on the desorption yield

The H_2- and CO desorption yields, calculated from equation 4.8, are shown in figure 6.1 for three different types of ions: Noble gas ions, hydrogen containing ions and oxygen containing ions.
The ions with an initial energy of 5keV impact on an Oxygen-Free High Conductivity (OFHC) copper sample.

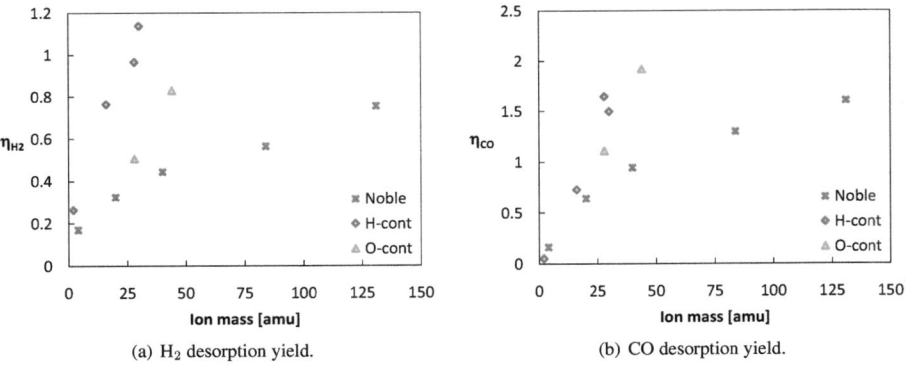

(a) H_2 desorption yield. (b) CO desorption yield.

Figure 6.1: H_2- and CO desorption yields of three different ion types incident on an OFHC-copper sample.

For hydrogen- and oxygen containing ions a steeper and almost linear increase of the desorption yield with the ion mass is visible in figure 6.1, while for noble gas ions a slower increase is observable. This behavior was observed for all applied ion energies between 1 and 7keV.

6.1.2 Energy and mass dependance of the desorption yield

For these three types of ions the following general characteristics have been observed within the considered energy range between 1 and 7keV:

- H_2 is the predominant desorbed gas for light ions, e.g. H_2^+, He^+ and CH_4^+.

- CO is the predominant desorbed gas for all other ions with higher masses.

- CO^+ and N_2^+-ions show a similar desorption behavior and are therefore handled together.

For the main desorbed gases, e.g. H_2, CO and CO_2, the desorption yields are presented in the following sections for the three different ion types in terms of an increasing energy and mass of the incident ions.

6.1.2.1 Noble gas ions

For noble gas ions the desorption yields of H_2 are increasing with the ion energy up to a maximum followed by a smooth decrease as it is shown in figure 6.2. For heavier ions this maximum is shifted to higher energies: He^+-ions show a maximum at around 1keV, Ne^+- and Ar^+-ions at around 5keV. Kr^+- and Xe^+-ions do not show a maximum within 1 and 5keV.

The H_2 desorption yields are continuously increasing with an increasing mass of the incident ions as it is shown in figure 6.3.

The CO desorption yields are increasing with an increasing energy and mass of the incident ions as it is shown in figure 6.4 and 6.5, respectively. An energy dependant maximum is reached for He^+-ions at 3keV and for Ne^+-ions at 5keV.

The CO_2 desorption yields are increasing with an increasing energy and mass of the incident ions as it is shown in figure 6.6 and 6.7, respectively. For He^+- and Ne^+-ions an energy dependant maximum is reached at around 3keV.

6.1. RESULTS

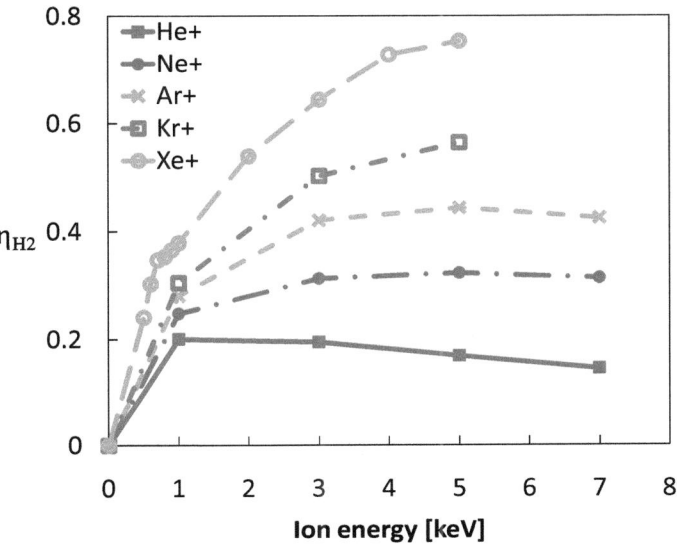

Figure 6.2: H_2 desorption yields of noble gas ions incident on copper as function of the ion energy.

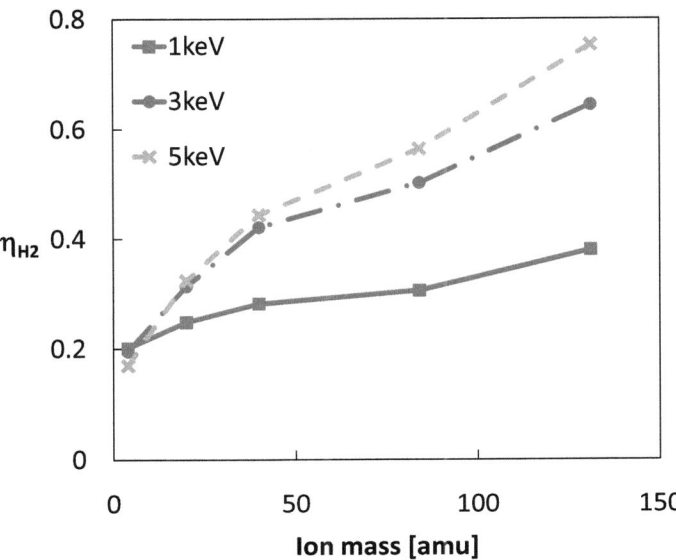

Figure 6.3: H_2 desorption yields of noble gas ions incident on copper as function of the ion mass.

CHAPTER 6. RESULTS AND DISCUSSION

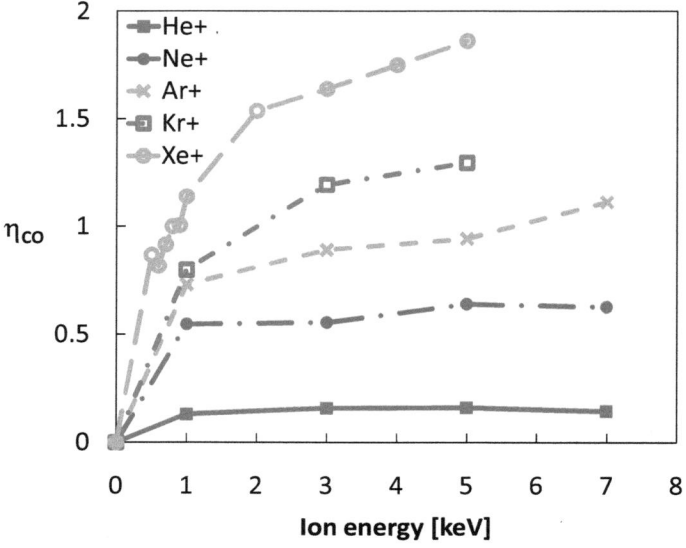

Figure 6.4: CO desorption yields of noble gas ions incident on copper as function of the ion energy.

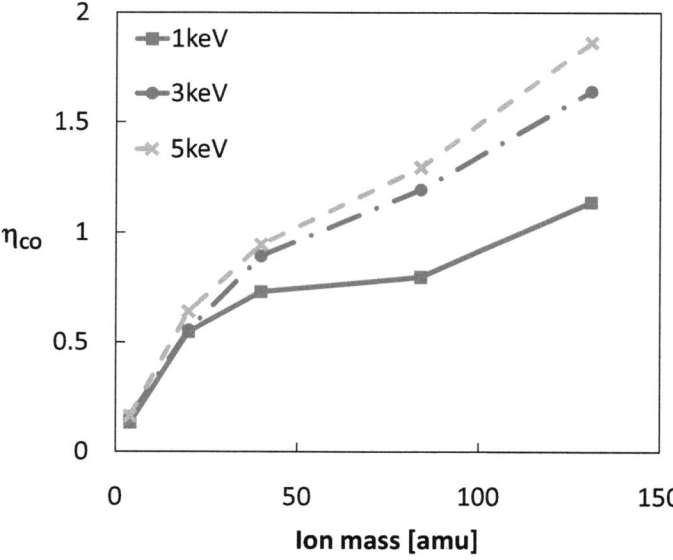

Figure 6.5: CO desorption yields of noble gas ions incident on copper as function of the ion mass.

6.1. RESULTS

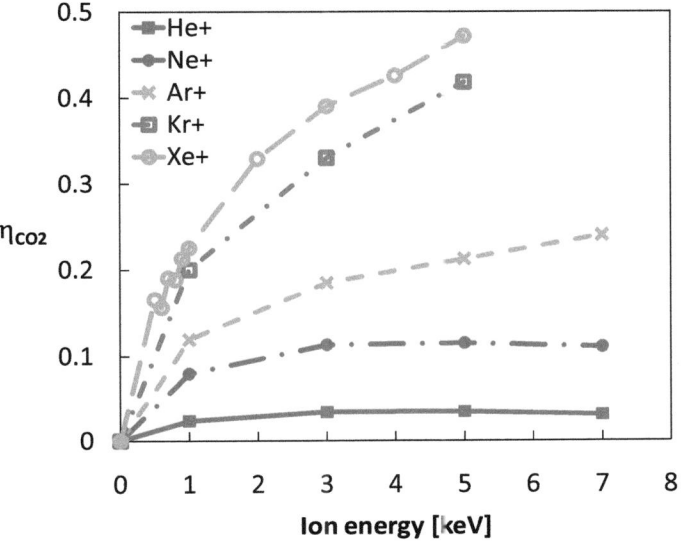

Figure 6.6: CO_2 desorption yields of noble gas ions incident on copper as function of the ion energy.

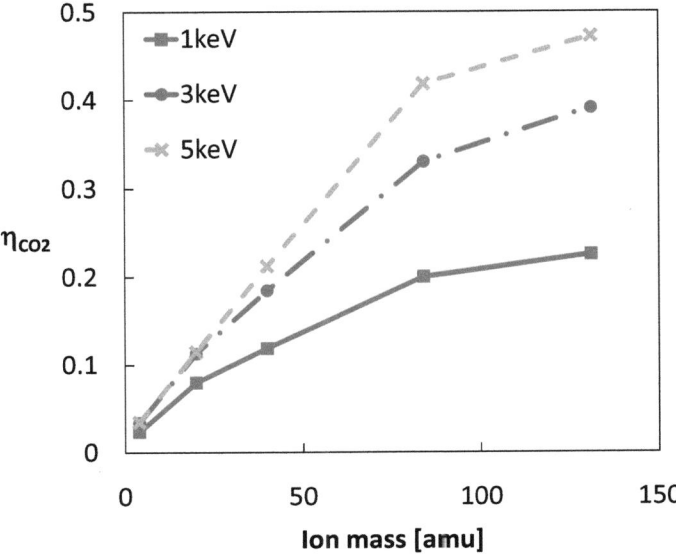

Figure 6.7: CO_2 desorption yields of noble gas ions incident on copper as function of the ion mass.

CHAPTER 6. RESULTS AND DISCUSSION

6.1.2.2 Hydrogen containing ions

In case of hydrogen containing ions the H_2 desorption yields show the tendency of saturation with higher ion energies which is visible in figure 6.8. The H_2 desorption yields are increasing continuously with the mass of the incident ions as it is shown in figure 6.9.

The desorption yields of CO and CO_2 show a similar behavior for hydrogen containing ions: They are increasing with the energy and mass of the incident ions as it is shown in figures 6.10 and 6.11, respectively for CO and in figures 6.12 and 6.13, respectively for CO_2. In case of H_2^+-ions an energy dependant maximum is reached at around 3keV for both desorption yields.

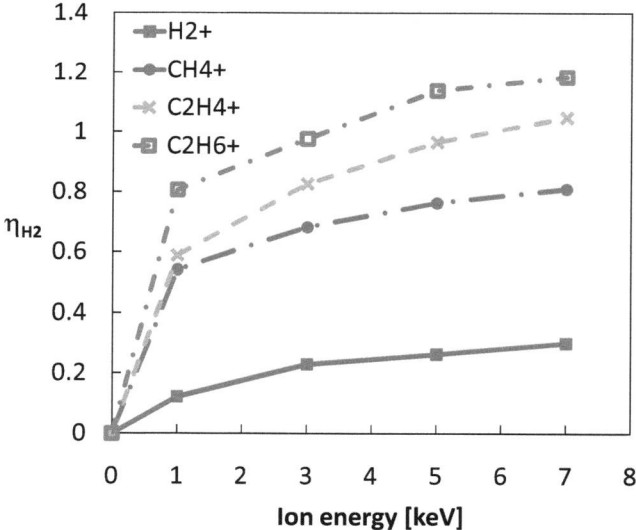

Figure 6.8: H_2 desorption yields of hydrogen containing ions incident on copper as function of the ion energy.

6.1. RESULTS

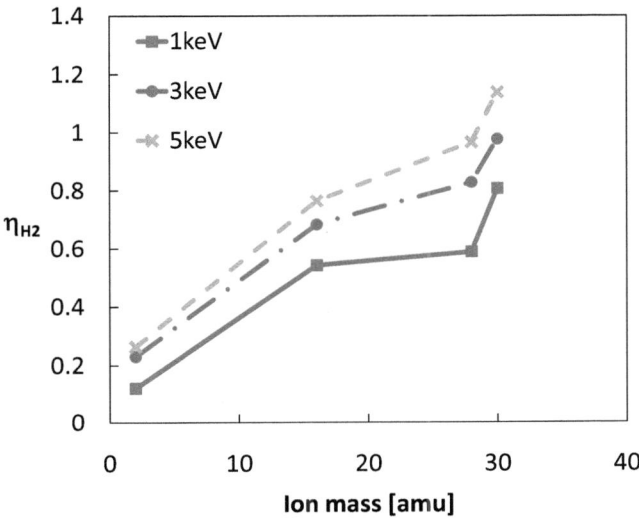

Figure 6.9: H_2 desorption yields of hydrogen containing ions incident on copper as function of the ion mass.

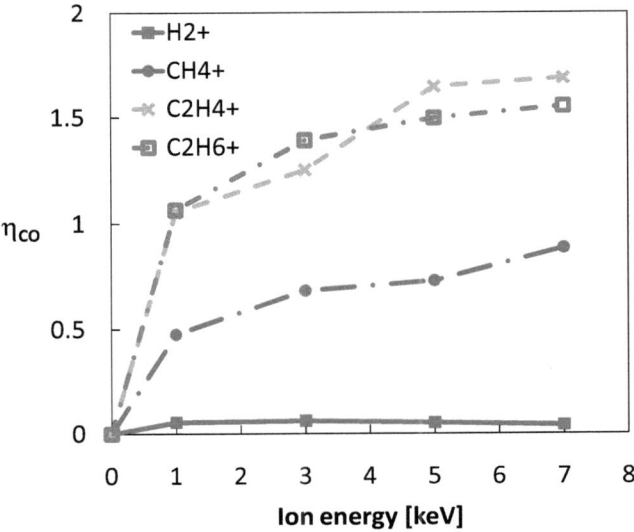

Figure 6.10: CO desorption yields of hydrogen containing ions incident on copper as function of the ion energy.

CHAPTER 6. RESULTS AND DISCUSSION

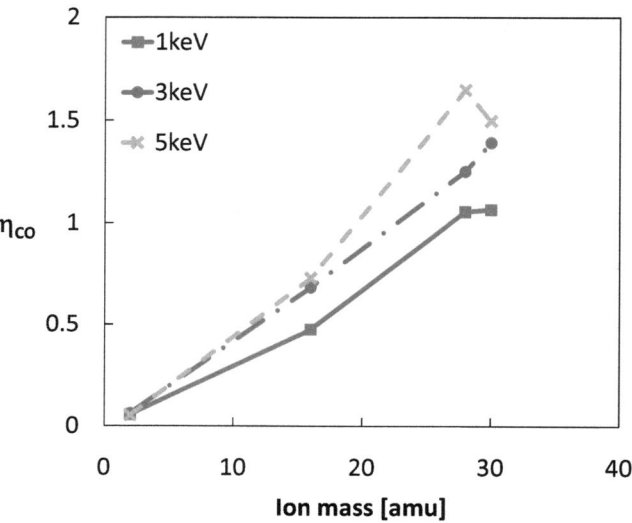

Figure 6.11: CO desorption yields of hydrogen containing ions incident on copper as function of the ion mass.

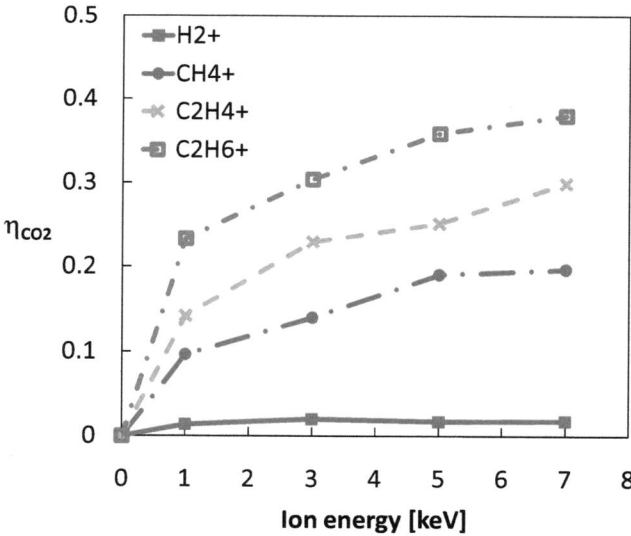

Figure 6.12: CO_2 desorption yields of hydrogen containing ions incident on copper as function of the ion energy.

6.1. RESULTS

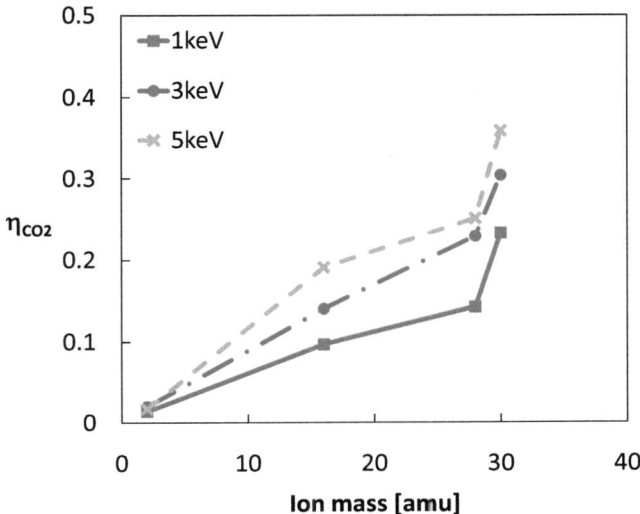

Figure 6.13: CO_2 desorption yields of hydrogen containing ions incident on copper as function of the ion mass.

6.1.2.3 Other type of ions: Oxygen containing ions and nitrogen ions

For this type of ions ions the desorption yields of H_2 are increasing with an increasing energy and mass of the incident ions as it is shown in figure 6.14 and 6.15, respectively. The H_2 desorption yields of N_2^+- and CO^+-ions are similar but an energy dependant maximum is reached at 5keV in the case of N_2^+-ions.

The desorption yields of CO and CO_2 for N_2^+- and CO^+-ions show a similar behavior: They are increasing with the energy and mass of the incident ions as it is shown in figures 6.16 and 6.17, respectively for the desorption yields of CO and in figures 6.18 and 6.19, respectively for the desorption yields of CO_2. Only the CO_2 desorption yields reach an energy dependant maximum at around 5keV.

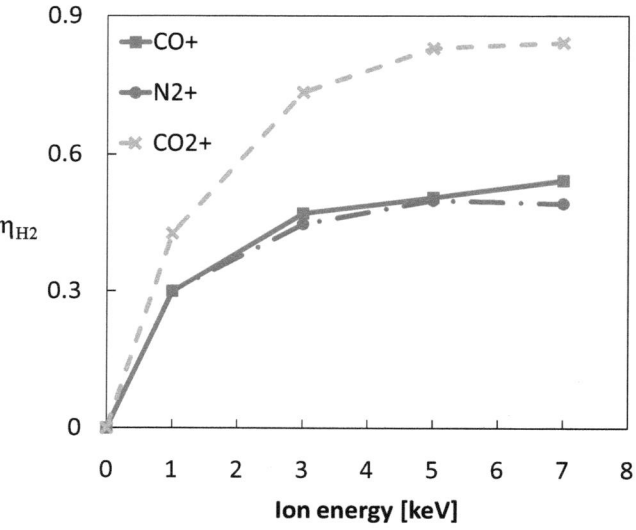

Figure 6.14: H_2 desorption yields of N_2^+-ions and oxygen containing ions incident on copper as function of the ion energy.

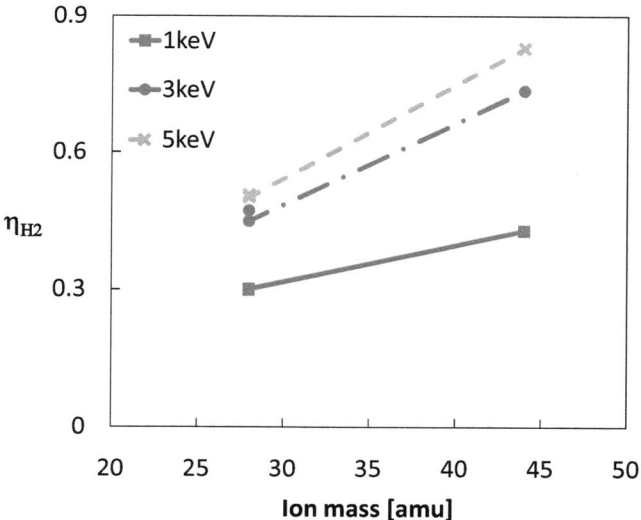

Figure 6.15: H_2 desorption yields of N_2^+-ions and oxygen containing ions incident on copper as function of the ion mass.

6.1. RESULTS

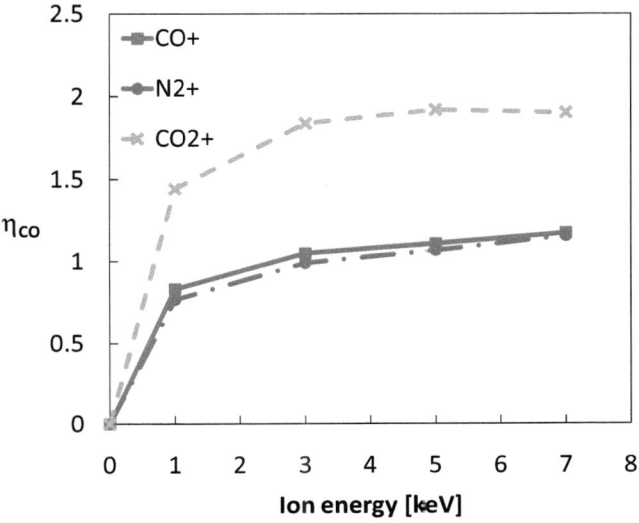

Figure 6.16: CO desorption yields of N_2^+-ions and oxygen containing ions incident on copper as function of the ion energy.

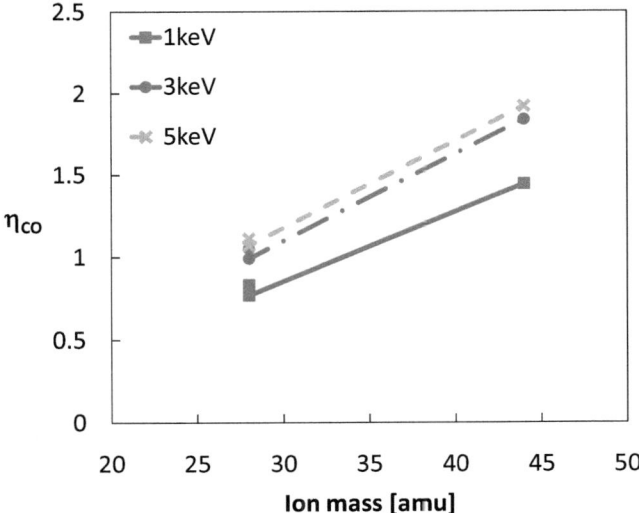

Figure 6.17: CO desorption yields of N_2^+-ions and oxygen containing ions incident on copper as function of the ion mass.

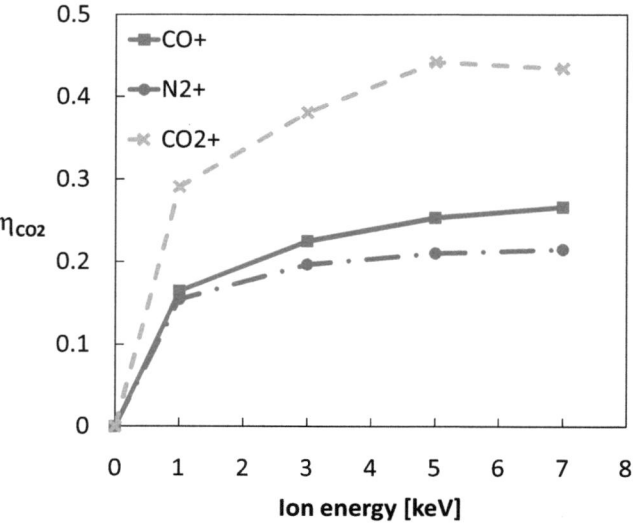

Figure 6.18: CO_2 desorption yields of N_2^+-ions and oxygen containing ions incident on copper as function of the ion energy.

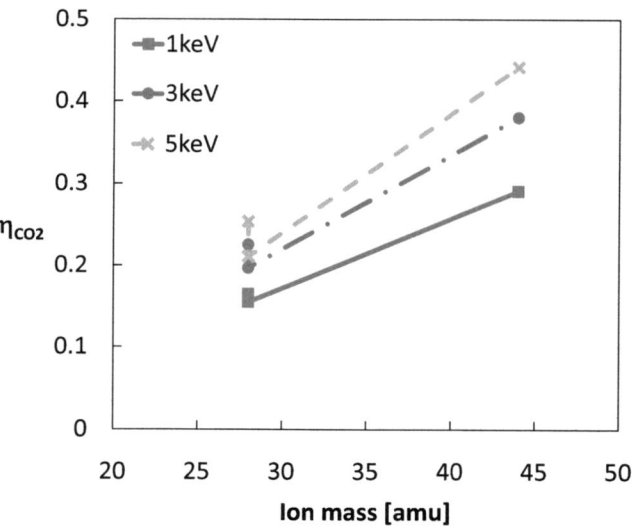

Figure 6.19: CO_2 desorption yields of N_2^+-ions and oxygen containing ions incident on copper as function of the ion mass.

6.1.3 Dose dependance of the desorption yield

6.1.3.1 Various ions incident on OFHC-copper

Dose dependent measurements of the desorption yields were carried out with 3 and 7keV ions. The behavior of the desorption yields versus ion dose were quite similar for these two energies. Additionally the RGA mass signal of the incident ion was recorded and the corresponding desorption yield was calculated.

The measurements which are presented in figures 6.20 to 6.25 show the desorption yields of 7keV ions incident on OFHC-copper. At an accumulated ion dose between 10^{14} and 10^{15}ions/cm^2 all desorption yields start to decrease due to an incident surface cleaning and saturate for doses $> 10^{16}$ions/cm^2. This decrease is stronger for heavier ions.

In figure 6.20 it is visible that at the beginning of irradiation the desorption yields of CO and CO_2 are slightly increasing in the case of He$^+$-ions and at an ion dose of 10^{15} ions/cm^2 the CO desorption yield starts to exceed the before predominating H_2 desorption yield (see also section 6.2).

The desorption yields of 7keV N_2^+- and CO$^+$-ions, both with mass 28, are shown in figure 6.24 and 6.25, respectively. In case of N_2^+-ions the contribution to mass 28 could be subtracted due to the separate RGA signal at mass 14 with its correlated sensitivity factor. Therefore the evolution of the pure CO desorption yield with the ion dose can be measured.

In case of N_2^+-ions the CO and CO_2 desorption yields are decreasing with the ion dose while the N_2 desorption yield (at mass 14) approaches unity. In contrary for CO$^+$-ions, after an initial decrease, an increase of the desorption yields of CO and CO_2 is visible which is stronger for the CO_2 desorption yields and exceeds their initial values (see also section 6.2).

CHAPTER 6. RESULTS AND DISCUSSION

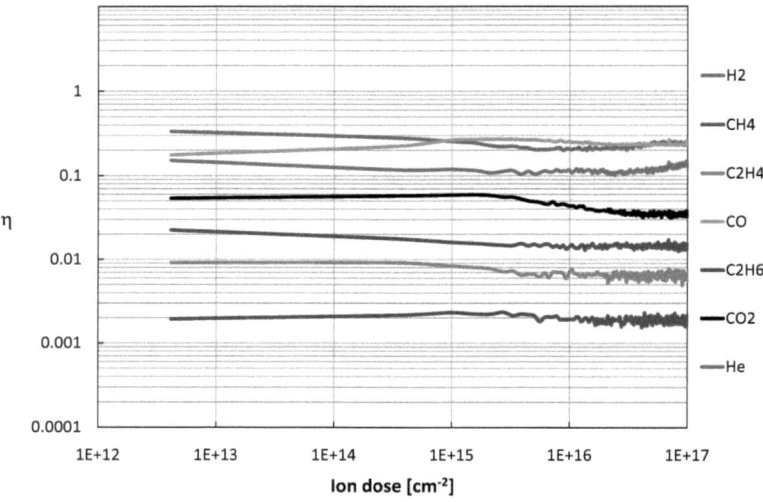

Figure 6.20: Dose dependance of 7keV He$^+$-ions incident on copper.

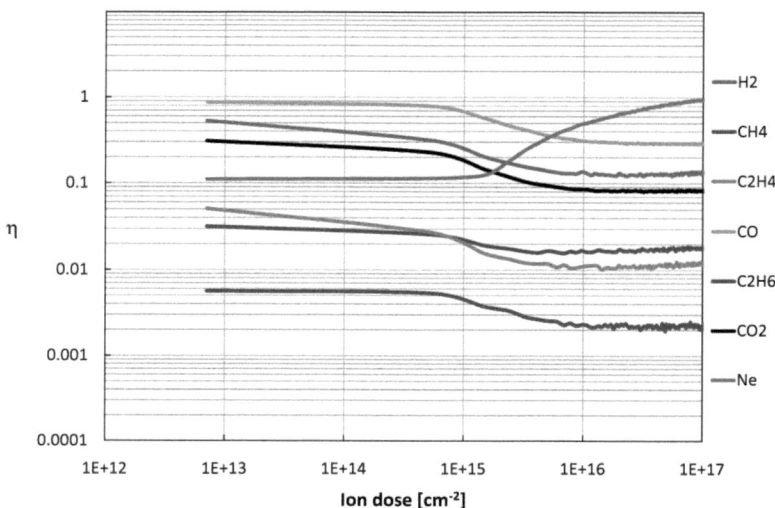

Figure 6.21: Dose dependance of 7keV Ne$^+$-ions incident on copper.

6.1. RESULTS

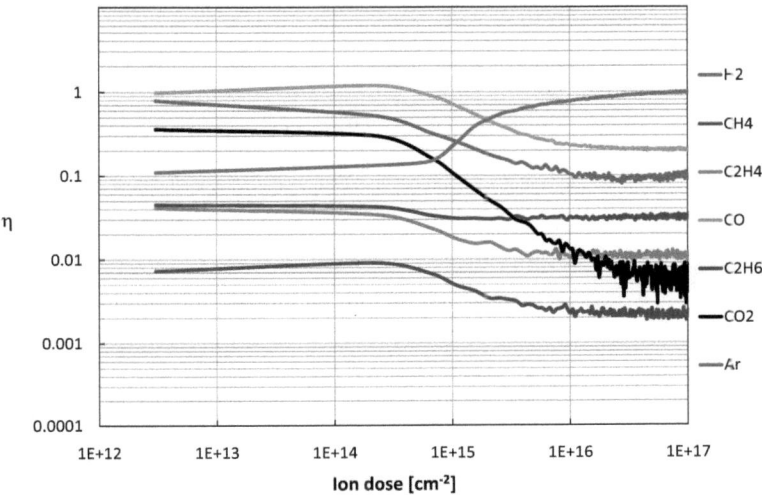

Figure 6.22: Dose dependance of 7keV Ar^+-ions incident on copper.

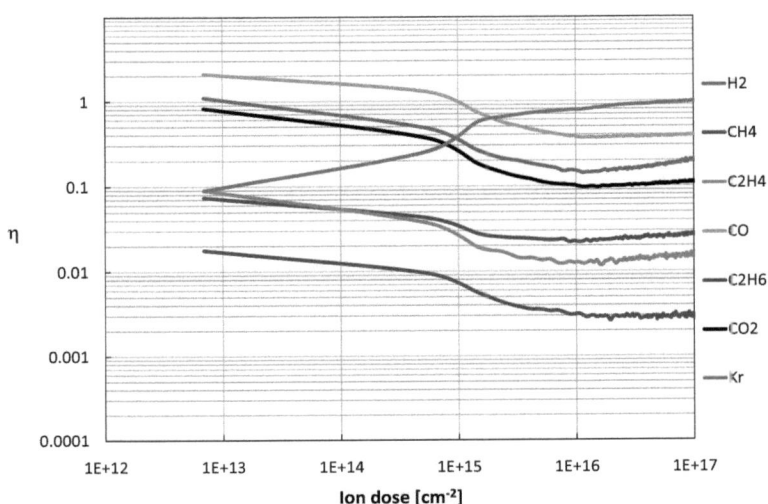

Figure 6.23: Dose dependance of 7keV Kr^+-ions incident on copper.

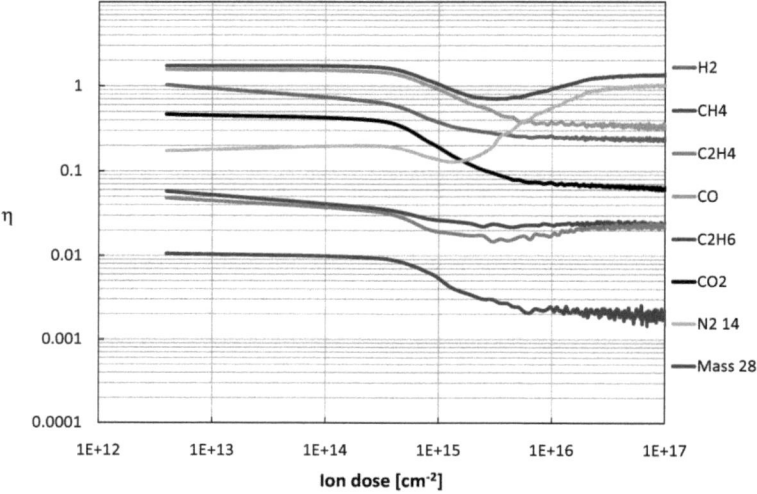

Figure 6.24: Dose dependance of 7keV N_2^+-ions incident on copper.

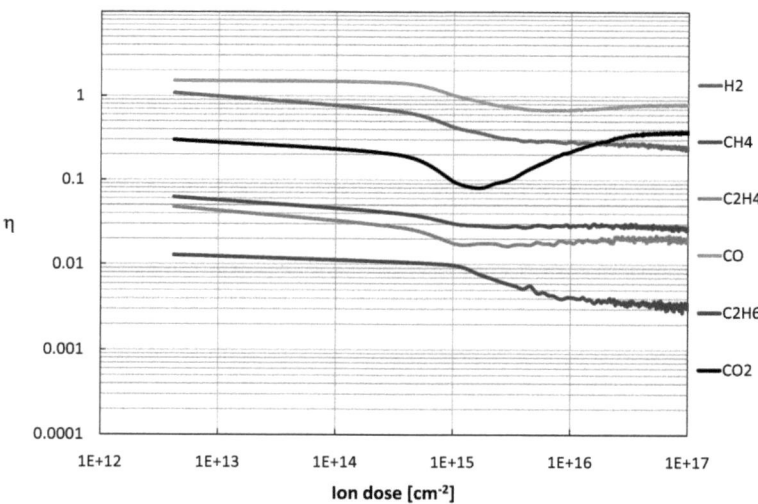

Figure 6.25: Dose dependance of 7keV CO^+-ions incident on copper.

6.1. RESULTS

6.1.3.2 Ar$^+$-ions incident on different target materials

The influence of the target material on the dose dependance of the desorption yield was investigated with 7keV Ar$^+$-ions incident on stainless steel[1] (figure 6.26), on copper coated stainless steel (figure 6.27), on silver coated stainless steel (figure 6.28), on gold coated stainless steel (figure 6.29) and on beam screen copper[2]. The behavior of the desorption yields with an increased ion dose is very similar to the one already mentioned above. In case of stainless steel it was observed that the desorption yields of H$_2$ and CH$_4$ start to saturate while all other desorption yields are still decreasing. In some cases like in figure 6.28 the desorption yields of CH$_4$ and C$_2$H$_4$, after an initial decrease, start to increase at ion doses of $\approx 10^{15}$ions/cm^2 (see also section 6.2).

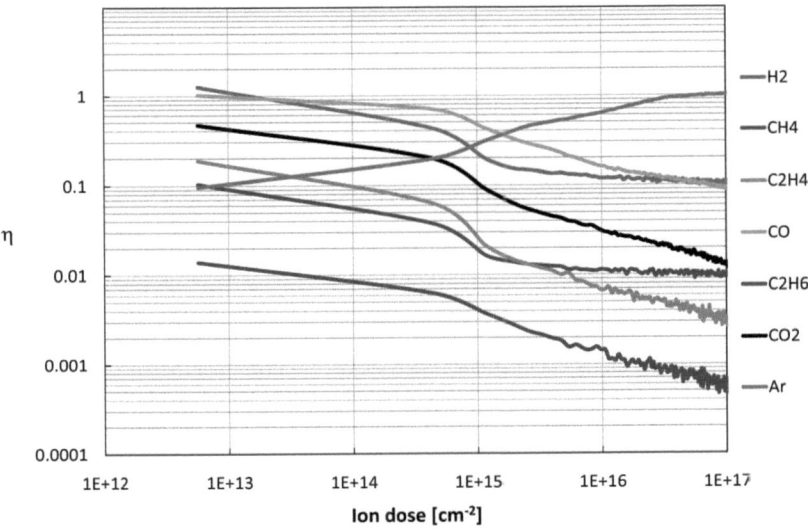

Figure 6.26: Dose dependance of 7keV Ar$^+$-ions incident on stainless steel.

---

[1] mainly consistent of Fe, Cr, Ni and Mo
[2] as used in the LHC

CHAPTER 6. RESULTS AND DISCUSSION

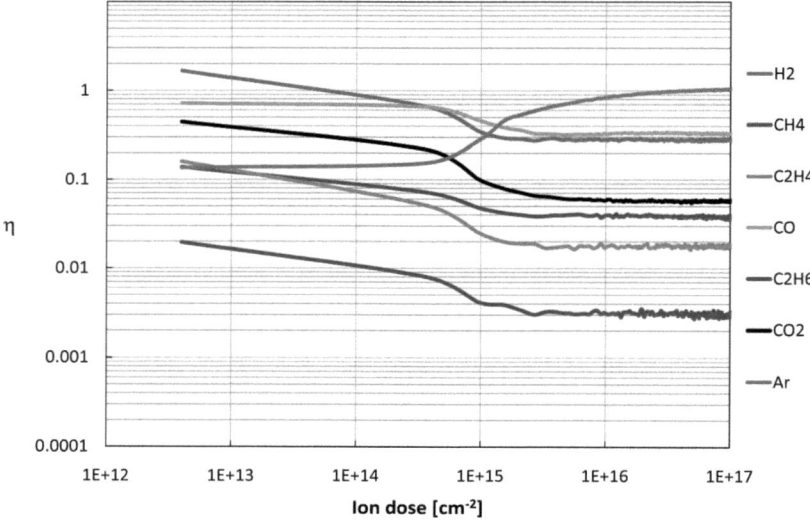

Figure 6.27: Dose dependance of 7keV Ar^+-ions incident on 1μm copper coated stainless steel.

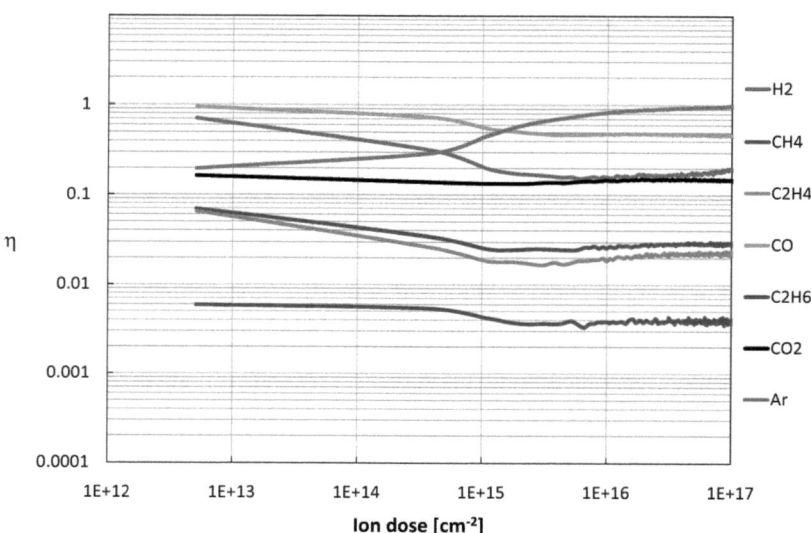

Figure 6.28: Dose dependance of 7keV Ar^+-ions incident on 1μm silver coated stainless steel.

6.1. RESULTS

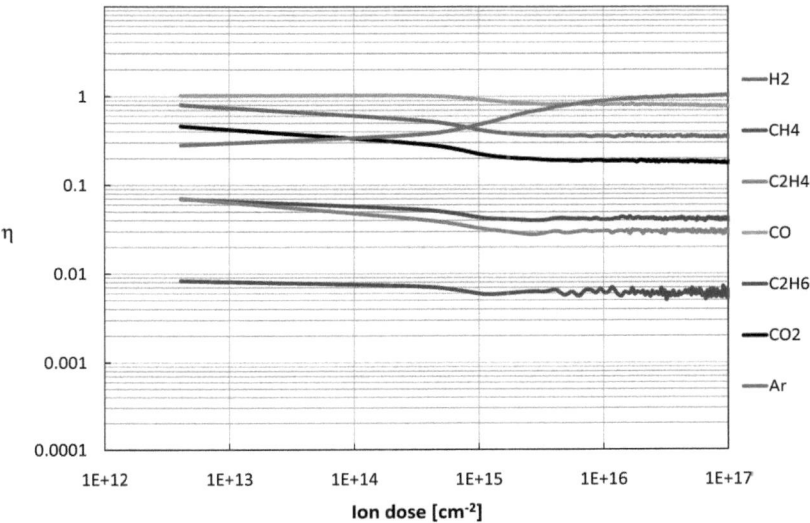

Figure 6.29: Dose dependance of 7keV Ar^+-ions incident on $1\mu m$ gold coated stainless steel.

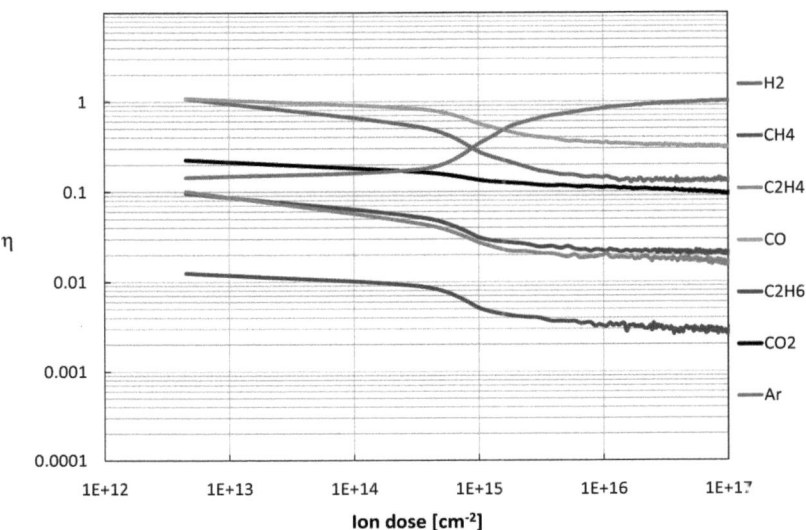

Figure 6.30: Dose dependance of 7keV Ar^+-ions incident on beam screen copper.

6.1.4 Desorption cross section and total coverage

According to equation 4.15 it is possible to calculate the desorption cross section σ and the total coverage N_0 from the dose dependant desorption yield measurements. Since the desorption yields saturate at high ion doses (see also section 6.2) it is not possible to fit all data points with this law, hence only initial desorption cross sections and coverage can be calculated. As an example figure 6.31 shows the fit result for the initial CO_2 desorption yields in the case of Ar^+-ions incident on copper.

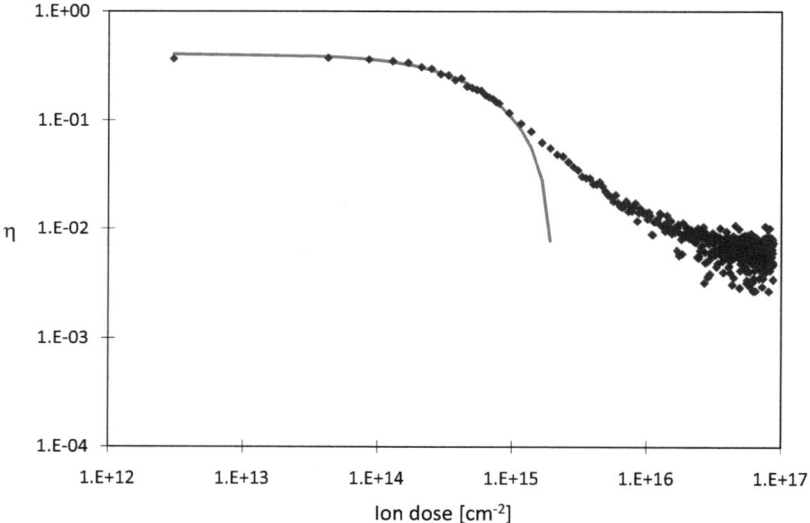

Figure 6.31: Fit result for the calculation of σ and N_0.

6.1.4.1 Noble gas ions incident on OFHC-copper

From figure 6.32 it is possible to see that there is no significant difference in the desorption cross section for 3 and 7keV noble gas ions. Further the cross sections are increasing with the ion mass. The magnitudes of the measured initial cross sections are in good agreement with the literature (cf. [112]). A compilation of cross sections can be found in chapter A.4. In case of He-ions it was not possible to calculate the desorption cross section and total coverage of CO and CO_2 due to the initial increase of the desorption yield with the ion dose.

6.1. RESULTS

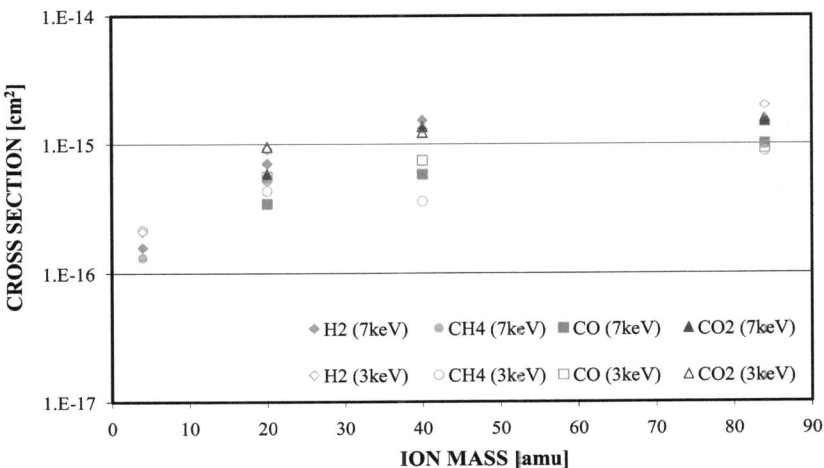

Figure 6.32: Desorption cross section σ obtained for 3 and 7keV noble gas ions incident on copper.

Figure 6.33 shows that the initial coverage of the sample is strongest for CO (coverage $\geq$ than one monolayer, see also section 6.2), followed by CO_2 and H_2 which are almost equal, followed by CH_4. 3 and 7keV incident ions show similar results for the initial coverage.

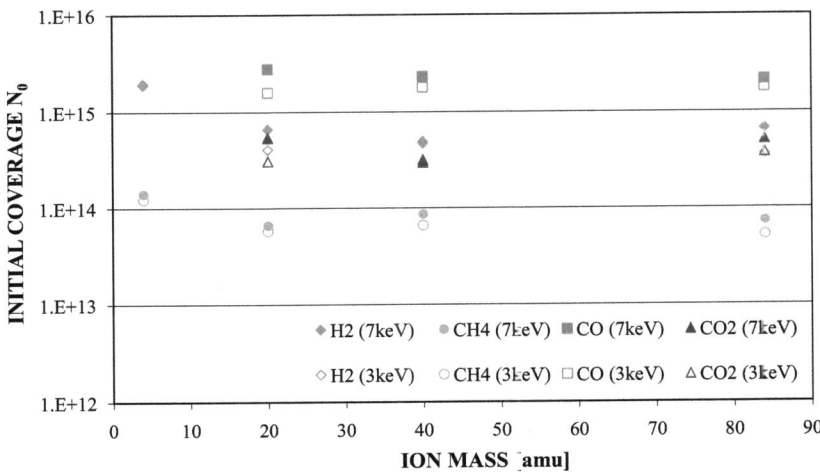

Figure 6.33: Initial coverage N_0 obtained for 3 and 7keV noble gas ions incident on copper.

6.1.4.2 Ar$^+$-ions incident on different target materials

The desorption cross sections of 7keV Ar$^+$-ions incident on different target materials are similar (cf. [19]) and are smallest for CO as it is shown in figure 6.34. The measured cross sections are again in good agreement with the literature (cf. [112]).

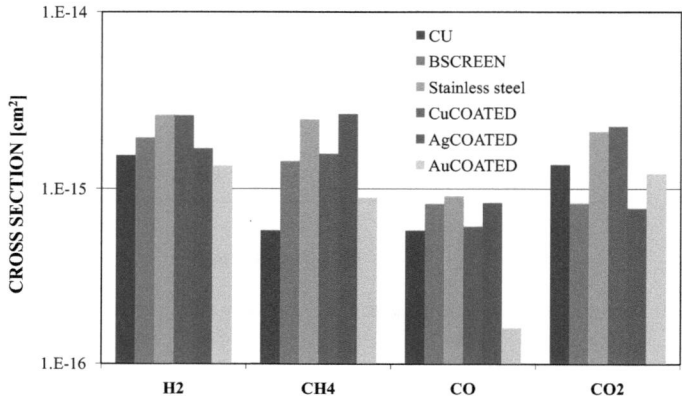

Figure 6.34: Desorption cross section obtained for 7keV Ar$^+$-ions incident on different target materials.

Also the initial coverage is similar for the different target materials. It is strongest for CO, followed by CO_2 and H_2 which are almost equal, followed by CH_4 as it is shown in figure 6.35.

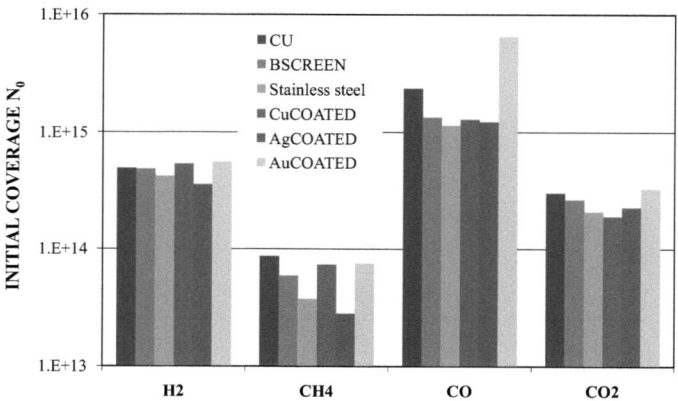

Figure 6.35: Initial coverage obtained for 7keV Ar$^+$-ions incident on different target materials.

6.2 Discussion

6.2.1 Correlation between sputter- and desorption yields

Figure 6.36 shows a comparison of calculated sputter yields (Y) for Ne^+-, Ar^+- and Xe^+-ions incident on copper from [100, 101] versus our measured desorption yields (η). Despite the fact that the two ordinates have different scalings, the two yields show similarities. Since the sputter yield can be calculated from the electronic and nuclear energy loss of the ions in matter as it was explained in chapter 4.6, also the obtained desorption yields were fitted with the energy loss of the ions in matter.

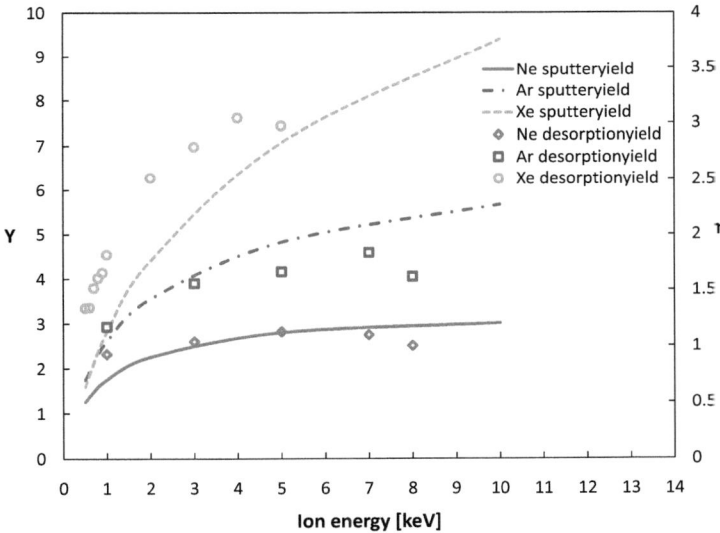

Figure 6.36: Calculated sputter yields (Y) versus measured desorption yields (η) for noble gas ions incident on copper.

The electronic and nuclear energy loss of the ions can be calculated with the SRIM programm [113]. For the case of molecular ions only an indirect calculation of the energy loss with SRIM is possible:

After the molecular ions with the mass m_{tot} penetrate the target, they are dissociated and create single-collision cascades [114, 115]. Hence it can be assumed that each of its components was having the same speed v before impact, given by the total kinetic energy E_{tot} (non relativistic):

$$v = \sqrt{\frac{2E_{tot}}{m_{tot}}} \qquad (6.1)$$

CHAPTER 6. RESULTS AND DISCUSSION

Hence the energy of each component E_c with the mass m_c is given by

$$E_c = \frac{m_c \cdot E_{tot}}{m_{tot}} \qquad (6.2)$$

With this energy the energy loss can be calculated with SRIM. The energy loss of the molecular ion is then assumed to be the sum over the energy losses of its components.

In figure 6.37 the ratio of electronic to nuclear energy loss is shown for H_2^+- and noble gas ions incident on copper. It shows that in contrary to heavy ions, where the electronic energy loss is almost negligible, the electronic energy loss of light ions like H_2^+ and He^+ can even exceed the nuclear energy loss by orders.

Hence for the following considerations the sum of electronic and nuclear energy loss, the so called total energy loss $S_{tot} = S_e + S_n$, was used. The assumed unit of the energy loss can easily be transformed into other units for ions incident on copper by multiplication with the corresponding factors from table A.1.

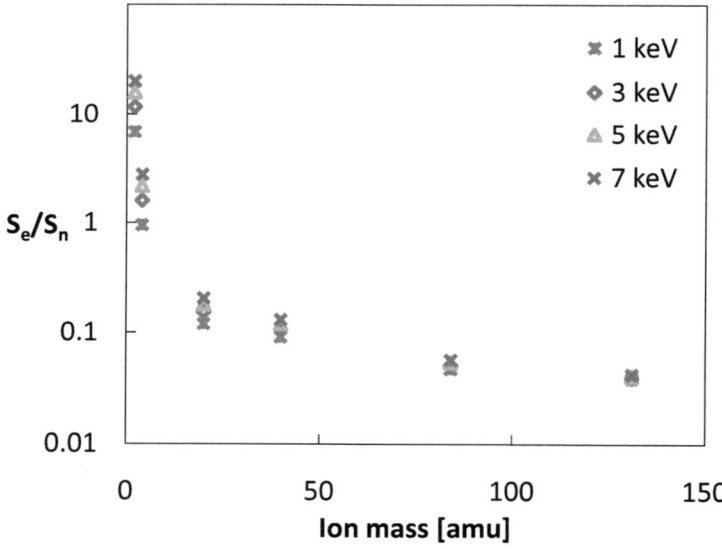

Figure 6.37: Ratio of electronic to nuclear energy loss for H_2^+- and noble gas ions incident on copper.

In this context also the different ion trajectories for light and heavy ions should be mentioned as it is shown in figure 6.38 on the example of SRIM simulations for He^+- and Kr^+-ions incident on copper at two different energies.

6.2. DISCUSSION

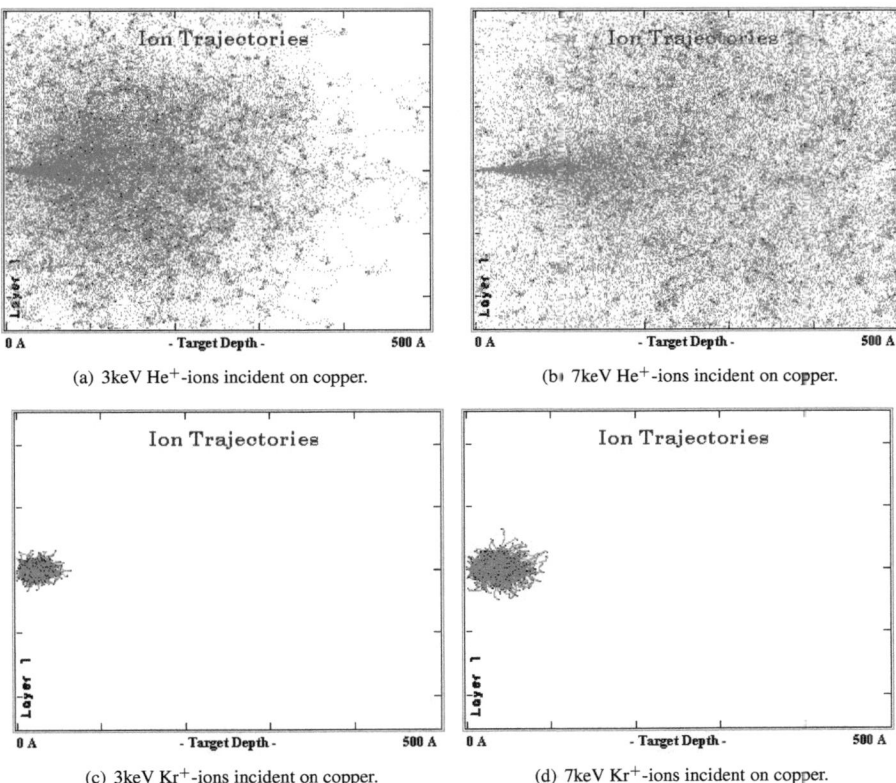

Figure 6.38: Ion trajectories of He$^+$- and Kr$^+$-ions incident on copper calculated with the SRIM program [113].

In relation to figure 6.38 our investigations[3] of an irradiated OFHC-copper sample with an electron microscope (magnification 1500x, 70° tilting angle) have shown different erosion of the sample surface due to the different mass of the incident ions as it is shown in figure 6.39 for accumulated ion doses of $\approx 6 \times 10^{17}$ ions/cm^2.

---
[3]figured out by Alexandre Gerardin - former TS/MME/MM section at CERN

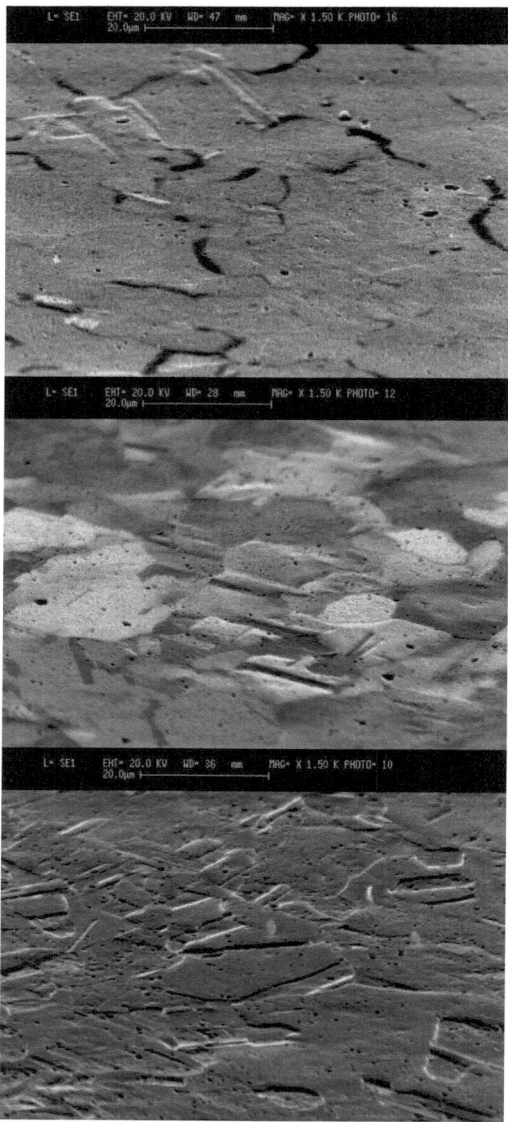

Figure 6.39: Microscopic view of an OFHC-copper sample. From top to bottom: Non irradiated, after He$^+$ irradiation, after Ar$^+$ irradiation, with an applied ion dose of $\approx 6 \times 10^{17}$ions/cm^2.

6.2. DISCUSSION

6.2.2 Different reasons for the desorption of molecules with different masses

The aim of this section is to show that the desorption of molecules with different masses can have its origin in the variable electronic and nuclear energy losses of the incident ions according to their masses, as it has been observed in our experiments for noble gas ions. Since the electronic energy loss is almost negligible for heavy ions but not for H_2^+- and He^+-ions, these two ions were used for the following investigations:

Equation 4.41, which is used to calculate sputter yields of noble gas ions (cf. [100, 101]), has been simplified to fit the electronic and nuclear energy loss with the desorption yields of H_2, CO and CO_2. The simplified equation is given by

$$\eta = k \cdot \frac{S_n(E)}{1 + AS_e(E)} \tag{6.3}$$

where the factor k serves as a proportionality factor (between energy loss- and desorption yield units) as well as a weighting factor for the nuclear energy loss $S_n(E)$ while the factor A acts only as a weighting factor for the electronic energy loss $S_e(E)$.
In case of He^+-ions the factor A differs from zero only for H_2 desorption.

	A	k
H_2	6,7	8,62
CO	0	5,43
CO_2	0	1,07

In case of H_2^+-ions the hydrogen desorption yield can only be fitted with the electronic energy loss itself. For the two other gases the results are

	A	k
CO	3,75	9,27
CO_2	0	2,31

It can be seen that the factor A is decreasing with an increasing mass of the desorbed molecules, hence the conclusion could be driven that for light desorbed molecules, e.g. H_2, mainly the electronic energy transfer to the lattice causes desorption while for heavier desorbed molecules it is more the direct nuclear momentum transfer between two particles. The fits for H_2 desorption for He^+-ions are shown in figure 6.40 and for H_2^+-ions in figure 6.41.

CHAPTER 6. RESULTS AND DISCUSSION

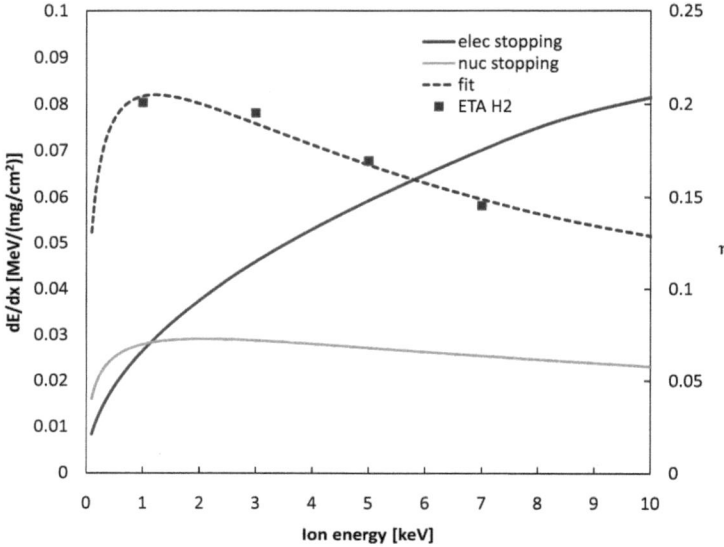

Figure 6.40: H_2 desorption yields of He^+-ions fitted with the electronic and nuclear energy loss.

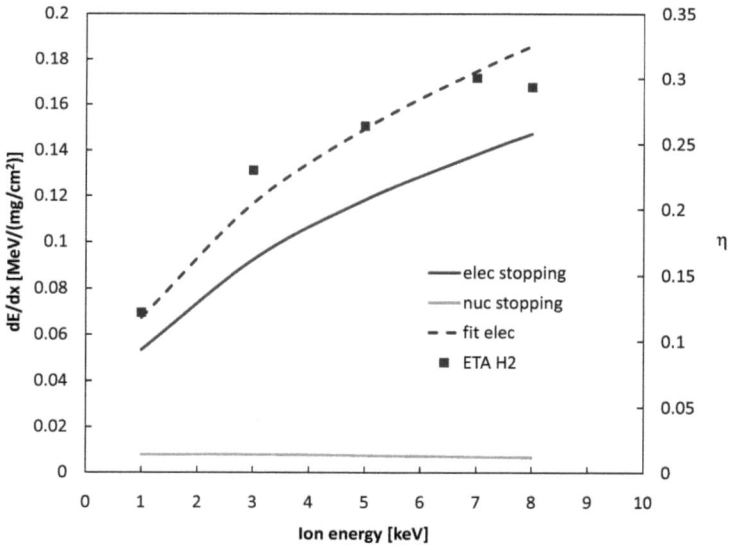

Figure 6.41: H_2 desorption yields of H_2^+-ions fitted only with the electronic energy loss.

6.2. DISCUSSION

6.2.3 Prediction of low energy ion desorption yields

In section 6.1.2 the results for the energy and mass dependance of the desorption yields have been shown. All these desorption yields can be fitted with the energy and mass dependant total energy loss of the ions in matter. Therefore by the help of calculations figured out with SRIM it is possible to predict the low energy ion desorption yields for the three mentioned ion types.

6.2.3.1 Noble gas ions

For noble gas ions the last square fits of the desorption yields of H_2, CO and CO_2 are shown in figures 6.42 to 6.44 together with the corresponding fit equations, given by

$$\eta = A \cdot S_{tot}^b \tag{6.4}$$

where A serves as a proportionality factor. The exponent b in the fit equation increases linearly with the mass m of the desorbed gas following equation 6.5 subjected to the condition that the hydrogen desorption yields for He^+-ions can be neglected (the motivation was discussed in the previous section):

$$b = 6,5 \cdot 10^{-3} m + 0,47 \tag{6.5}$$

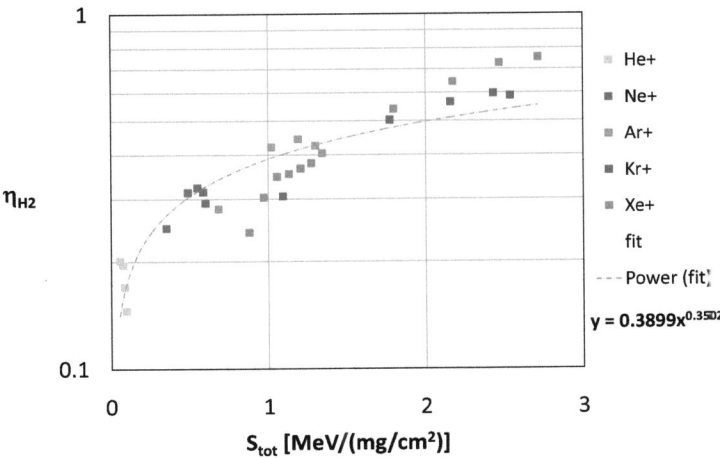

Figure 6.42: H_2 desorption yields as a function of the total energy loss obtained for noble gas ions incident on copper.

CHAPTER 6. RESULTS AND DISCUSSION

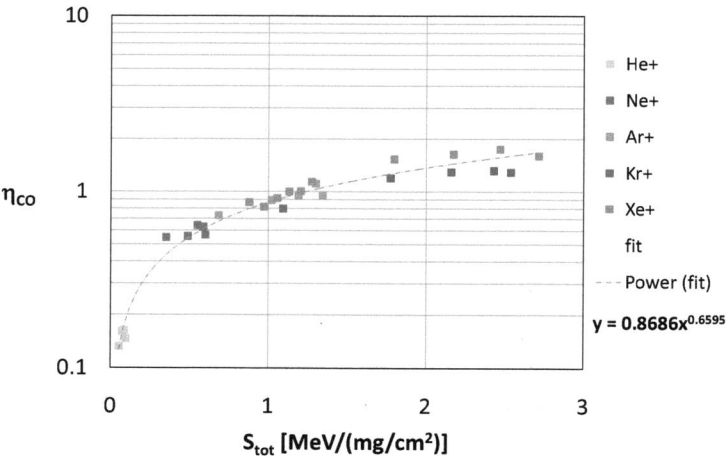

Figure 6.43: CO desorption yields as a function of the total energy loss obtained for noble gas ions incident on copper.

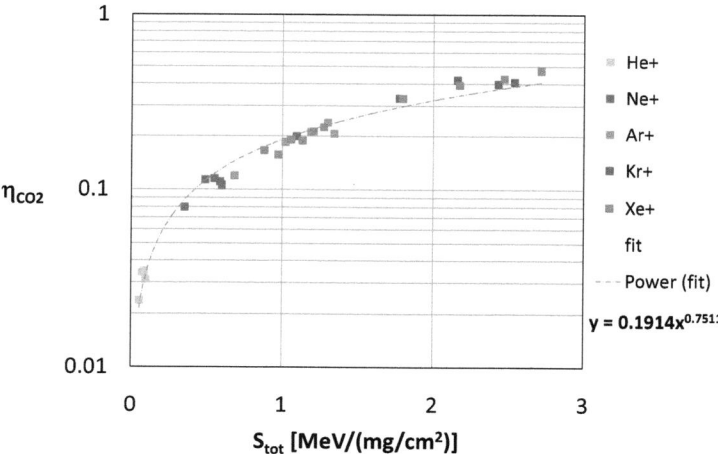

Figure 6.44: CO_2 desorption yields as a function of the total energy loss obtained for noble gas ions incident on copper.

6.2.3.2 Hydrogen containing ions

For hydrogen containing ions the last square fits of the desorption yields of H_2, CO and CO_2 are shown in figures 6.42 to 6.44 together with the corresponding fit equations. No linear mass dependance of

6.2. DISCUSSION

the exponent from the desorbed gas was found here.

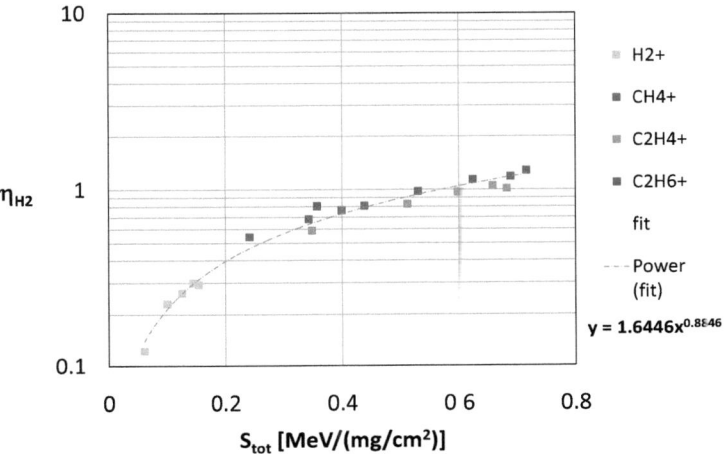

Figure 6.45: H_2 desorption yields as a function of the total energy loss obtained for hydrogen containing ions incident on copper.

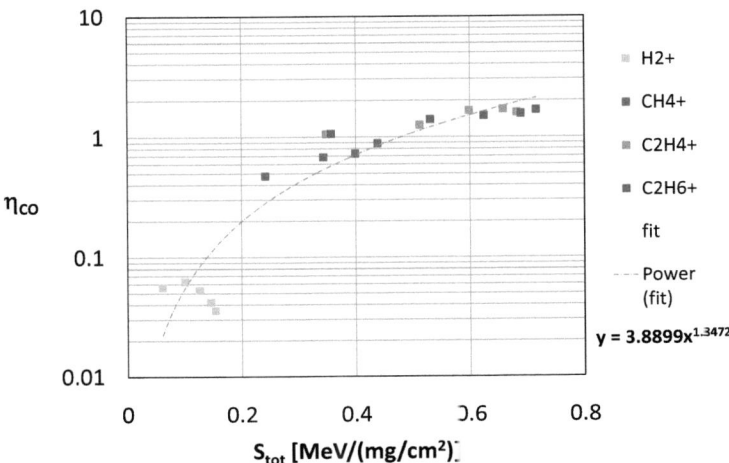

Figure 6.46: CO desorption yields as a function of the total energy loss obtained for hydrogen containing ions incident on copper.

CHAPTER 6. RESULTS AND DISCUSSION

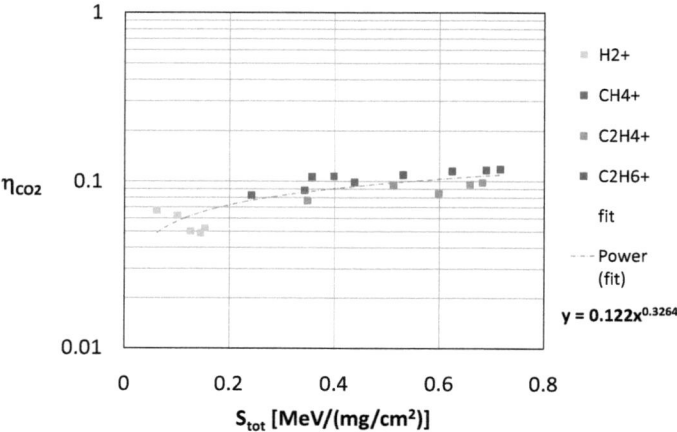

Figure 6.47: CO_2 desorption yields as a function of the total energy loss obtained for hydrogen containing ions incident on copper.

6.2.3.3 Other type of ions: Oxygen containing ions and nitrogen ions

For this type of ions the last square fits of the desorption yields of H_2, CO and CO_2 are shown in figure 6.42 to 6.44 together with the corresponding fit equations. The exponent in the fit equation is approx. one for all desorbed gases.

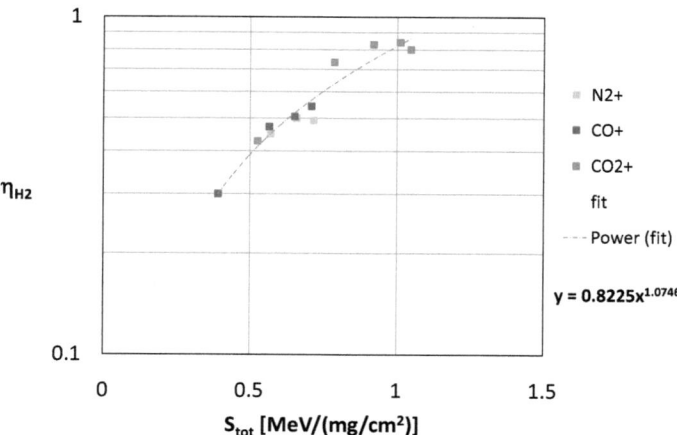

Figure 6.48: H_2 desorption yields as a function of the total energy loss obtained for N_2^+-ions and oxygen containing ions incident on copper.

6.2. DISCUSSION

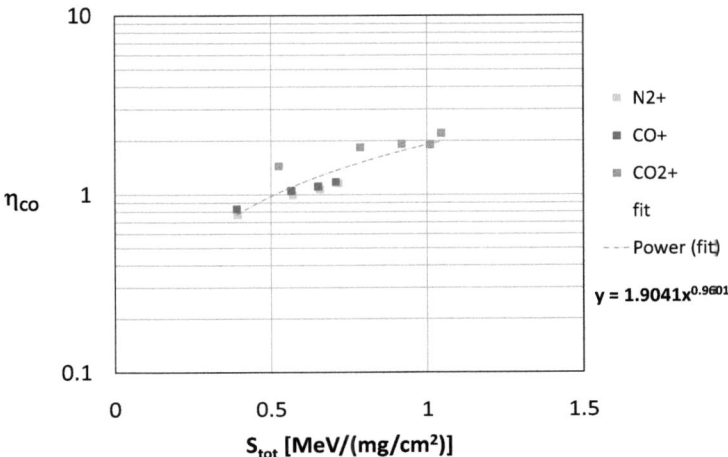

Figure 6.49: CO desorption yields as a function of the total energy loss obtained for N_2^+-ions and oxygen containing ions incident on copper.

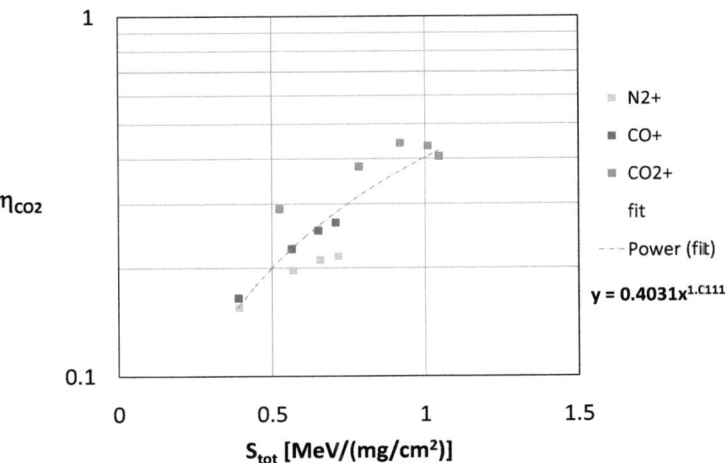

Figure 6.50: CO_2 desorption yields as a function of the total energy loss obtained for N_2^+-ions and oxygen containing ions incident on copper.

6.2.4 Influence of the beam shape on the desorption yield

Since the beam was not uniform in its ion density, investigations have been done to calculate this influence on the desorption yield. Therefore a Gauss distribution of the ion density J (A/cm^2) as a function of the beam radius r was assumed which is given by

$$J = J_0 \cdot e^{-ar^2} \tag{6.6}$$

The factor a controls the width of the "bell" in the Gauss function as it is shown in figure 6.51.

To calculate the desorption yield of the total beam, the beam area was subdivided into small slices with the width dr (see figure 6.51).

The number of incoming ions N_i on each slice is given by

$$N_i = \bar{j}_i \cdot A_i \tag{6.7}$$

where $j = J/e$ is the number of incoming ions/area (e is the elementary charge), $\bar{j}_i$ is the average ion density between $j(r)$ and $j(r+dr)$ and $A_i = \pi \cdot [(r+dr)^2 - r^2]$ is the area of one slice i. The total number of incoming ions is then given by the sum over N_i.

Using equations 4.9 and 4.12 for the calculation of the desorption yield η at the time t, the number of particles desorbed N_d from each slice is given by

$$N_{d,i} = \bar{\eta}_{p,i} \cdot N_i \tag{6.8}$$

where $\bar{\eta}_p$ is the average desorption yield between $\eta_p(r)$ and $\eta_p(r+dr)$. The total number of desorbed particles is given by the sum over $N_{d,i}$.

Hence the desorption yield η of the total beam area at the time t is given by

$$\eta(t) = \frac{\Sigma N_{d,i}}{\Sigma N_i} \tag{6.9}$$

Figure 6.52 shows the desorption yield of the total beam area as a function of the ion dose D given by $D = t \cdot \Sigma N_i$, for the following parameters: $\sigma = 2 \times 10^{-15}$cm^2, $N_0 = 1 \times 10^{15}$cm^{-2} and $J_0 = 1 \times 10^{-7}$A/cm^2. It shows the appearance of a shoulder which has been observed in our experiments too but it does not show a saturation of the desorption yield with an increasing ion dose.

6.2. DISCUSSION

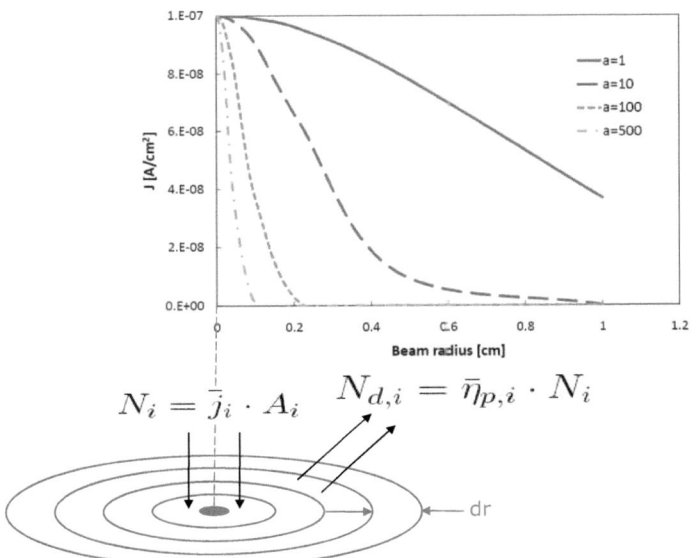

Figure 6.51: Subdivision of the beam area into small slices with an assumed Gauss ion density distribution.

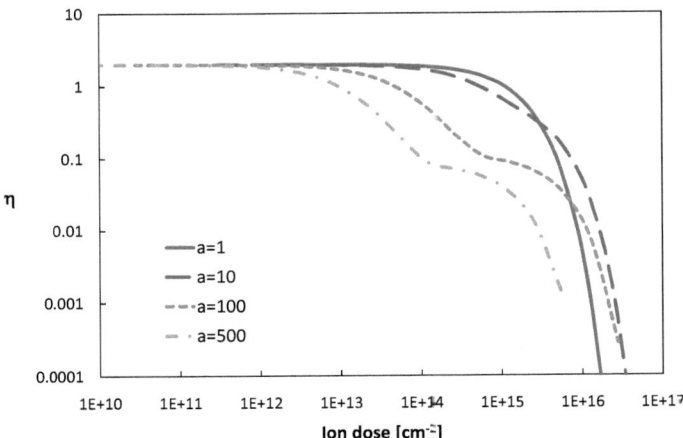

Figure 6.52: Total desorption yield of a beam with a Gauss ion density distribution as function of the ion dose.

6.2.5 Other effects during desorption

In [116] it is reported that the sputter yield of a Ruthenium single crystal decreases until steady state is reached at a primary-ion fluence of 2×10^{15} Ar$^+$/cm^2. Higher fluences up to $\approx 10^{17}$ have no further influence upon the sputtered Ru yield. The authors assume that two types of sites for sputtering exist, one which represents an undamaged site and one which has been damaged by a primary-ion impact. In [117] the creation of nano hillocks was observed which would also confirm this model of two "microscopic" contributions to the sputter (desorption) yield.

Surface vacancies, created by atoms removed by sputtering, are most probably responsible for the depression of the atomic Ru yield. The magnitude of the decrease, however, is significantly less for a contaminated surface compared to a clean one [116].

Most of the sputtered atoms originate from the topmost atomic layer as it was experimentally proved in [118] for a Ru substrate on copper in case of 3,6keV Ar$^+$-ions, hence the chemisorption of an impurity could lead to a reduction of the surface binding energy and, therefore, lead to an increase in the Ru sputtering yield. For example, the presence of carbon reduces the effective surface concentration of Ru, therefore, also lowering the Ru signal. Once the carbon impurity has been sputtered away, Ru is the sole constituent and its sputtering yield increases [116].

Such an increase has been observed in our experiments too, mostly pronounced for the desorption yields of CH_4 and C_2H_4 as it can be seen for example in figure 6.24.

The influence of oxide layers on the sputter yield was studied in [119] for two different copper oxide layers such as CuO and Cu_2O under Ar$^+$-ion impact at two different energies of 3 and 5keV. On samples which were kept in air for some days before introduction into the ultra-high vacuum chamber, copper has been identified only as Cu_2O on the surface. It was observed that during sputtering the CuO was quickly reduced to Cu_2O which itself was more stable against sputtering. However, Cu_2O could also be reduced – at least partially – to elemental copper using higher ion energies and doses. The oxide layer thickness of these prepared samples was more than 1000Å.

This value can be assumed as an upper bound for the thickness of the oxide layer of the samples used in our experiments since they were exposed to air only for some hours after their cleaning treatment.

According to [120] the sputter yields of a clean copper surface change at least by one order of magnitude when oxygen was admitted. The formation of two different layers (an oxygen adsorption layer and an oxide layer) as it was observed for other materials during oxygen admission could not be verified for copper [121]. In order to clean the samples from the oxide layer and other impurities according to [10] doses of $1,6 \times 10^{19}$ Ar$^+$-ions/cm^2 are required which is by far higher than the applied ion doses in our experiments.

In [10] it is mentioned that the major surface impurity after a solvent cleaning procedure is C ($\simeq 80$ at%) which can be 4-5 monolayer thick [122]. As it was already mentioned before, the incoming

6.2. DISCUSSION

ions dissociate after their impact on the surface (cf. [114, 115]). A comparison of the results for N_2^+- and CO^+ ions (see figures 6.25 and 6.25, respectively), both with mass 28, shows that for oxygen containing ions the effect of oxygen is obviously to oxidize C forming CO and CO_2, respectively (cf. [10]) while according to figure 6.25 the formation of CO_2 seems to be more efficient. To remove the carbon completely from the surface in [10] a argon/ 10% oxygen glow discharge is recommended.

6.2.6 Estimation of the measurement accuracy

A very detailed description of systematic measurement errors and error propagation during the calibration of the vacuum gauges in a similar experimental setup can be found in [123]. A total error for the calculation of the desorption yield was estimated in [123, 124] with 30%. This would mean a parallel shift by $\pm 15\%$ for the obtained desorption yields of our measurements.

Another, more unpredictable random error arises for the energy dependent measurements of the desorption yield because here the beam had to be readjusted for each ion energy (cf. chapter 5.5.1). Therefore a slightly change in the beam shape is possible and hence un-irradiated sample areas can attribute to the total desorption yield.

Nevertheless, the fact that the desorption yield of the incoming ion species approaches one with an increased ion dose (e.g. figure 6.22) states for a good calibration of the system:
Considering that the beam ions are implanted or trapped as interstitials at a certain depth depending on their energy, all vacancies will be filled up with incoming particles by time and hence the number of implanted particles N_{imp} in equation 6.10 goes to zero.

$$N_{inc} = N_{imp} + N_{des} \qquad (6.10)$$

At this stage the number of incident particles N_{inc} must be equal to the number of desorbed particles N_{des}. Since $N_{des} = \eta \cdot N_{inc}$ the desorption yield must go to one with progressing irradiation time. This behavior was observed for different ions during our measurements (cf. figures 6.21 to 6.23), hence the total error of the desorption yield can be indicated with approximately 10% and in addition this effect could be used to recalibrate the system since the ratio of the relative sensitivity factors of the RGA stays more or less constant.

Finally it should be mentioned here that the implantation of ions with identical energies into the surface is limited. In addition the desorption cross section is increasing with the ion mass and a formation of heavy adsorbates on the surface, e.g. CO_2, takes place. These effects lead to a pressure increase in the system and hence a strong pumping is required to ensure a certain lifetime of a circulating proton beam.

Chapter 7

Conclusion

This work was carried out in the vacuum group of the former AT department at CERN. Its aim was to investigate the low energy ion-induced desorption at room temperature in general and its effects on technical surfaces which are used in the LHC beam vacuum system in particular.

In this work the desorption yields of various ions have been studied on OFHC-copper as a function of the incident ion mass and energy. Due to their different desorption behavior, a classification into three different types of ions could be made which are: Noble gas ions, hydrogen containing ions and oxygen containing ions.

The measured desorption yields have shown similarities with the sputter behavior of single crystals. Therefore a model, which is used to calculate sputter yields from the energy loss of ions in matter, was successfully applied for the calculation of desorption yields.
Measured desorption cross sections for various ions incident on different target materials are in good agreement with the literature and have shown that they are increasing with the mass of the incident ions. Hence the composition of the residual gas for a circulating proton beam is changing from light to heavy ions by time. Additionally a formation of heavy adsorbates on the surface due to the interaction of the incoming ions with surface contaminations takes place. Together with the fact that the implantation of ions with identical energies into the surface is limited, these effects lead to a pressure increase in the system and hence a strong pumping is required to ensure a certain lifetime of the circulating proton beam.
Results have shown, that the initial coverage for the main gases, e.g. H_2, CH_4, CO and CO_2 on before cleaned samples is in the order of one monolayer and is not changing significantly for different materials. Hence the applied solvent cleaning procedure of the samples is not sufficient to reduce the initial desorption yield.

Appendix A

Appendix

A.1 Copper cleaning procedure

All samples were cleaned by Marina Malabaila[1]. The copper samples were treated in the same way than the beam screen of the LHC. The methodology of this cleaning procedure is listed bellow:

- **Chemical degreasing with detergent and ultrasonic**

    Formulation and operating parameters:

    - Detergent NGL 17.40 spec. ALU III: 10g/l.
    - Temperature: 50 - 60°C.
    - Time: 30 - 60 minutes.

- **Rinsing with water**

- **Pickling**

    Formulation and operating parameters:

    - Hydrochloric acid : 50%.
    - Temperature: 20°C.
    - Time: 10 - 30 seconds.

- **Rinsing with water**

- **Passivation**

    Formulation and operating parameters:

    - Chromic acid 70 - 80g/l + sulphuric acid 3ml/l.
    - Temperature: 20°C.

---

[1] former TS/MME/CCS section at CERN

– Time: 10 - 20 seconds.

- **Rinsing with water**

- **Rinsing with demineralized water and alcohol**

- **Drying with clean compressed air and bake-out**

A.2 Bake-out procedure

After an exchange of samples or due to a power cut a bake-out of the system was necessary to reach a base-pressure in the low 10^{-10} mbar range within a reasonable time. The bake-out temperature ranges between 250 and 300°C depending on the respective parts of the system. However, the main UHV-chamber is baked at 250°C for 24 hours. The heat-up and cool-down rate is 50°C per hour for all heaters.

At the end of each bake-out the gauges were degassed. For this purpose the cool-down was stopped at 200°C and a 30 minutes long degassing of the Bayard-Alpert gauges followed. After a further cool down the RGA was degassed at 150°C for 30 minutes.

A.3 Stopping units

Multiplication factor	Stopping units
89,197	eV/Å
891,97	keV/μm
891,97	MeV/mm
1	keV/(ug/cm^2)
1	MeV/(mg/cm^2)
1000	keV/(mg/cm^2)
105,52	eV/(1×10^{15} atoms/cm^2)
0,97991	L.S.S. reduced units

Table A.1: Multiplication factors for the stopping units of different ions incident on copper.

A.4 Compilation of ion impact desorption cross-section

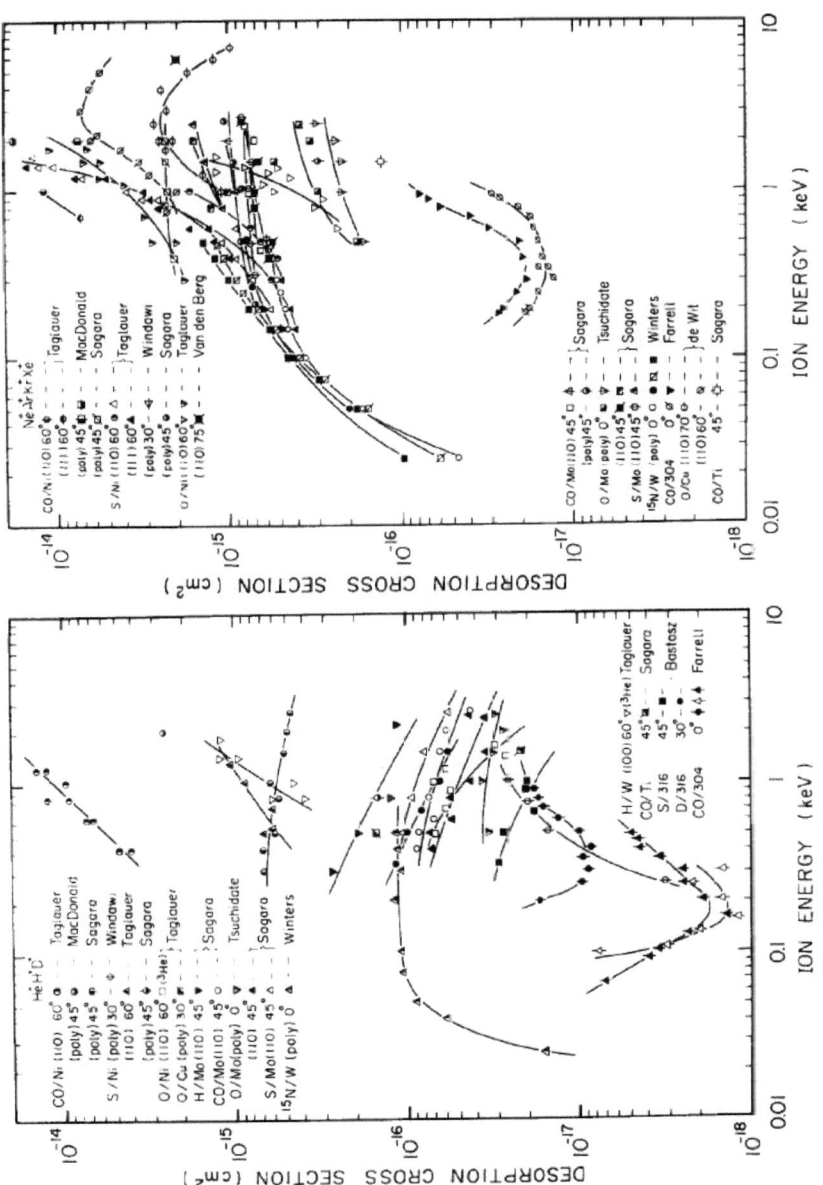

Figure A.1: Compilation of ion impact desorption cross section from [112].

A.5 Source electronics

All electrodes are powered by an electronic board which contains different modules. This modules are "off the shelf" components in order to minimize the design time of the power supply. A detailed overview of the power supply components can be found in figure A.2. For the sake of completeness a picture of the finalized board can be found in figure A.5.

High voltage The electronic board is floated above ground up to 10kV. The high voltage itself is generated by a laboratory power supply and is applied on the grid electrode. In order to provide 230V AC to the mains of the board, an isolation transformer should be used. Since such a transformer was not available and was also quite expensive to buy, two 230V/6kV transformers mounted up-side down were used. To avoid damage to the electronic board due to high voltage sparking inside the vacuum chamber (cf. section 5.1.1.2), all the power supply outputs are protected with Transil diodes and gas discharge tubes.

Filament The filament is heated by a DC voltage from a power supply module. The emission current which returns to the filament through a shunt resistor ($1k\Omega$) and 2 Zener diodes, generates a DC voltage of about 50V from GND. As the grid electrode is around 200V above GND, the voltage between grid and filament is about 150V. The voltage across the shunt passes through a voltage follower and is provided to an error amplifier to control the heating of the filament. The error amplifier has a capacitor in the feedback to avoid high frequency oscillations and a Zener diode at the output to limit the voltage across the filament. A detailed overview of the components is shown in figure A.3.

Grid and extraction The extraction voltage is generated by a DC/DC module. The module's output is not able to reach 0V, therefore its reference is directly connected to the grid electrode to respond to the requirements (in fact the voltage for the grid module was risen to set the reference for the extraction part). The grid voltage itself is generated by the help of an external series transistor. A detailed overview of the components is shown in figure A.4.

Repeller The repeller voltage arrives from the biased part of the filament and can be selected with a jumper to 0V or 50V.

A.5. SOURCE ELECTRONICS

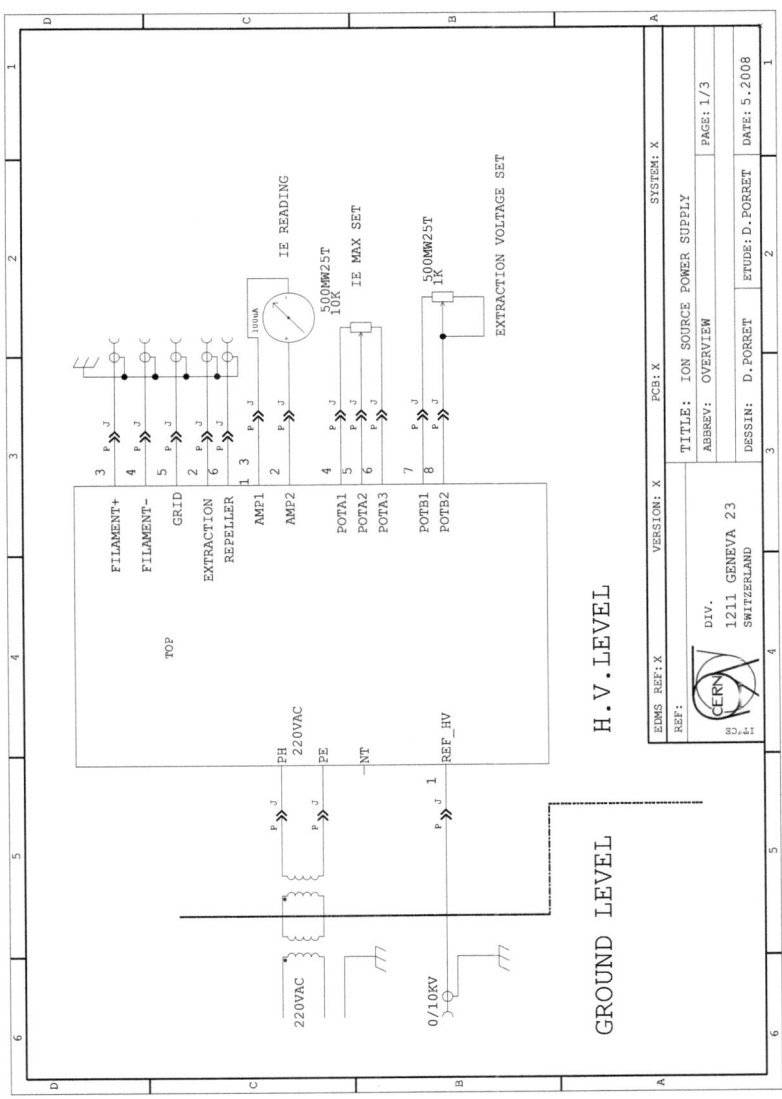

Figure A.2: Overview of the board components.

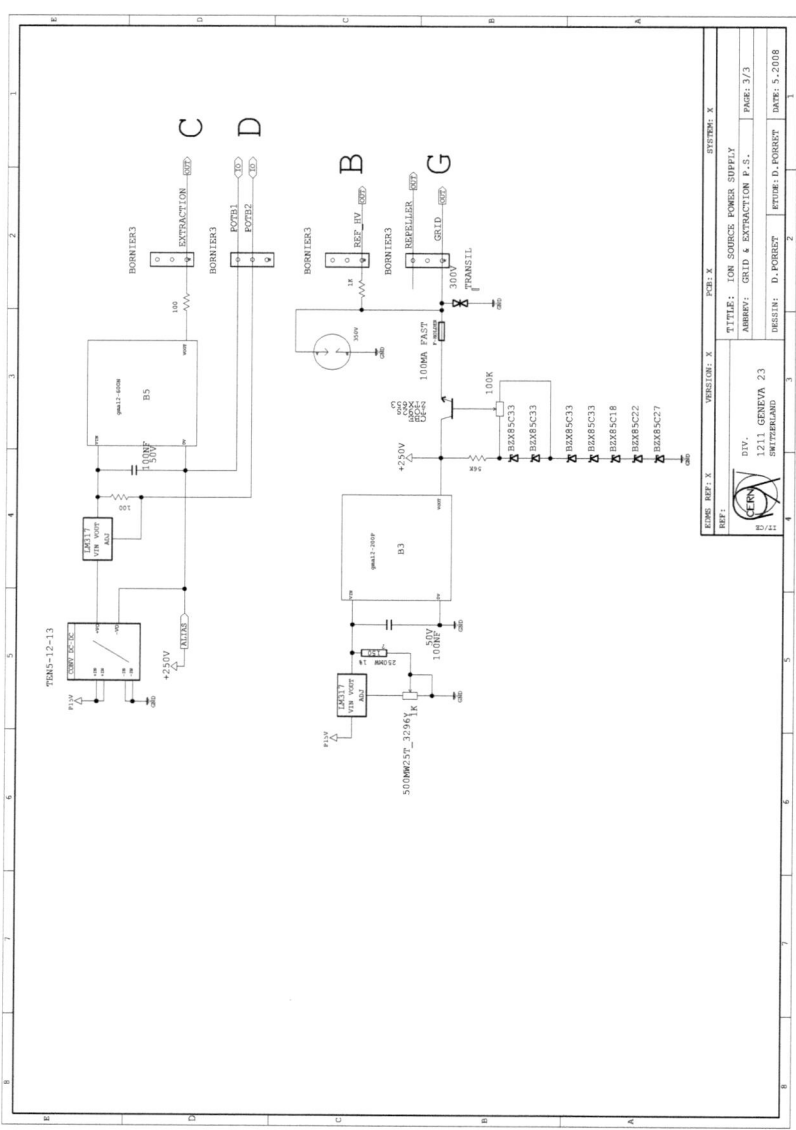

Figure A.3: Board components of grid and extraction.

A.5. SOURCE ELECTRONICS

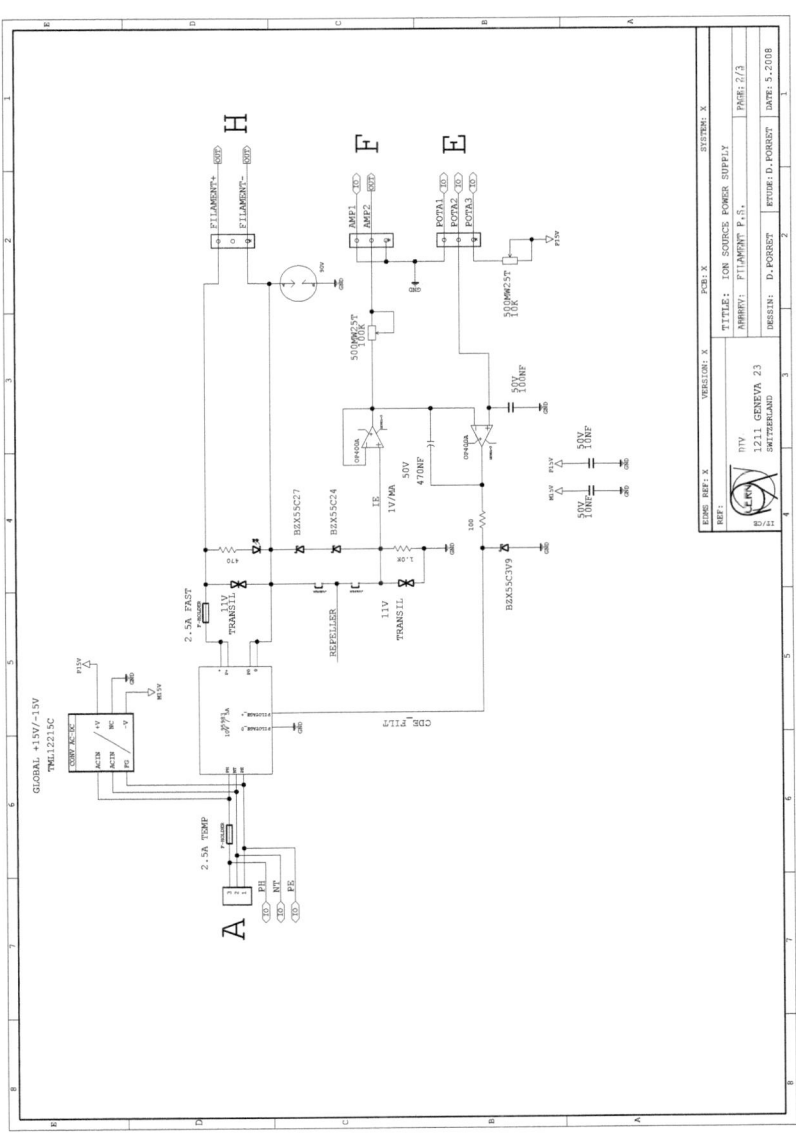

Figure A.4: Board components of the filament.

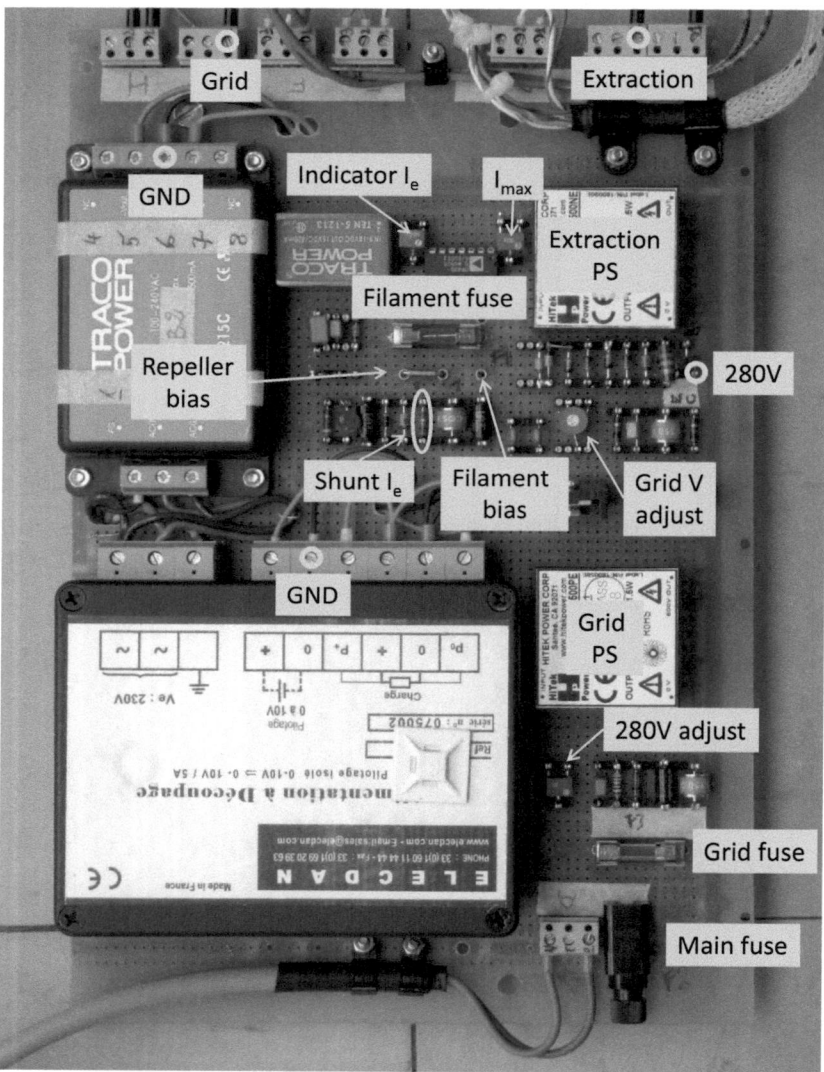

Figure A.5: Picture of the finalized board.

Bibliography

[1] Kingdon, K. H. and Langmuir, I. The Removal of Thorium from the Surface of a Thoriated Tungsten Filament by Positive Ion Bombardment. *Phys. Rev.*, 22(2):148–160, 1923.

[2] Calder, R. S. Ion induced gas desorption problems in the ISR. *Vacuum*, 24(10):437–443, 1974.

[3] McCracken, G. M. and Stott, P. E. Plasma surface reactions in Tokamaks. *NuclearFusion (Review Paper)*, 19(7):889–981, 1979.

[4] Horikoshi, G. and Kobayashi, M. A simple understanding of net outgassing rate as a function of pumping speed. *Journal of Vacuum Science and Technology*, 18(3):1009–1012, 1981.

[5] Winters, H. F. and Sigmund, P. Sputtering of chemisorbed gas (nitrogen on tungsten) by low-energy ions. *Journal of Applied Physics*, 45(11):4760–4766, 1974.

[6] Sigmund, Peter. Theory of Sputtering. I. Sputtering Yield of Amorphous and Polycrystalline Targets. *Phys. Rev.*, 184(2):383–416, 1969.

[7] Taglauer, E. and Marin, G. and Heiland, W. and Beitat, U. Study of the sputtering of adsorbates by low energy ions (O on Ni). *Surface Science*, 63:507–513, 1977.

[8] Taglauer, E. and Heiland, W. and Beitat, U. The influence of adsorption energies on ion impact desorption of surface layers. *Surface Science*, 89(1-3):710–717, 1979.

[9] Achard, M. H. and Calder, R. and Mathewson, A. G. The effect of bakeout temperature on the electron and ion induced gas desorption coefficients of some technological materials. *Vacuum*, 29:53–65, 1978.

[10] Mathewson, A. G. The surface cleanliness of 316 L + N stainless steel studied by SIMS and AES. *Vacuum*, 24(10):505–509, 1974.

[11] Dylla, H. F. A review of the wall problem and conditioning techniques for tokamaks. *Journal of Nuclear Materials*, 93-94(1):61–74, 1980.

[12] Waelbroeck, F. and Winter, J. and Wienhold, P. Cleaning and conditioning of the walls of plasma devices by glow discharges in hydrogen. *Journal of Vacuum Science & Technology A: Vacuum, Surfaces, and Films*, 2(4):1521–1536, 1984.

[13] Govier, R. P. and McCracken, G. M. Gas Discharge Cleaning of Vacuum Surfaces. *Journal of Vacuum Science and Technology*, 7(5):552–556, 1970.

[14] Lambert, R. M. and Comrie, C. M. A convenient electrical discharge method for eliminating hydrocarbon contamination from stainless steel UHV systems. *Journal of Vacuum Science and Technology*, 11(2):530–531, 1974.

[15] Ulrickson, M. and Dylla, H. F. and LaMarche, P. H. and Buchenauer, D. Particle balance in the Tokamak Fusion Test Reactor. *Journal of Vacuum Science & Technology A: Vacuum, Surfaces, and Films*, 6(3):2001–2003, 1988.

[16] Fischer E. Two Kilometers at 10^{-10} Torr. The CERN Intersecting Storage Rings for Protons. *Journal of Vacuum Science and Technology*, 9(4):1203–1208, 1972.

[17] Liu, S. M. and Rodgers, W. E. and Knuth, E. L. Interactions of hyperthermal atomic beams with solid surfaces. *The Journal of Chemical Physics*, 61(3):902–904, 1974.

[18] Lozano, M. P. Ion-induced desorption yield measurements from copper and aluminium. *Vacuum*, 67(3-4):339–345, 2002.

[19] Mathewson, A. G. Ion induced desorption coefficients for titanium alloy, pure aluminium and stainless steel. Technical Report CERN-ISR-VA-76-5, CERN, Geneva, Mar 1976.

[20] Hilleret, N. *Fourth International Conference On Solid Surfaces Proc., Cannes, France*, 2:1221, 1980.

[21] Bender, M. *Untersuchung der Mechanismen schwerioneninduzierter Desorption an beschleunigerrelevanten Materialien*. PhD thesis, Johann Wolfgang Goethe-Universität, Frankfurt am Main, 2008. Presented on 22 Feb 2008.

[22] Mahner, E. Review of heavy-ion induced desorption studies for particle accelerators. *Phys. Rev. ST Accel. Beams*, 11(10):104801, 2008.

[23] CERN – European Organization for Nuclear Research, Route de Meyrin, CH-1211 Genève 23, Switzerland. http://www.cern.ch, April 2008.

[24] Wikipedia – The Free Encyclopedia. http://www.wikipedia.org, April 2008.

[25] Pettersson, T. S. and Lefèvre, P. The Large Hadron Collider: Conceptual Design. Technical Report CERN-AC-95-05 LHC, CERN, Geneva, Oct 1995.

[26] Evans, L. R. The Large Hadron Collider Project. *CERN-LHC-Project-Report-53*, Sep 1996.

[27] ATLAS collaboration. *ATLAS: technical proposal for a general-purpose pp experiment at the Large Hadron Collider at CERN*. LHC Tech. Proposal CERN-LHCC-94-43. CERN, Geneva, 1994.

BIBLIOGRAPHY

[28] Coughlan, G. D. and Dodd, J. E. *The ideas of particle physics – An introduction for scientists*. Cambridge Univ. Press, Cambridge, 1991.

[29] Novaes, S. F. Standard Model: An Introduction. http://www.citebase.org/abstract?id=oai:arXiv.org:hep-ph/0001283, 2000.

[30] CMS collaboration. *CMS, the Compact Muon Solenoid : technical proposal*. LHC Tech. Proposal CERN-LHCC-94-38. CERN, Geneva, 1994.

[31] ALICE collaboration. *ALICE: Technical proposal for a Large Ion collider Experiment at the CERN LHC*. LHC Tech. Proposal CERN-LHCC-95-71. CERN, Geneva, 1995.

[32] LHCb collaboration. *LHCb : Technical Proposal*. LHC Tech. Proposal CERN-LHCC-98-004. CERN, Geneva, 1998.

[33] Lebrun, Ph. Advanced Superconducting Technology for Global Science: The Large Hadron Collider at CERN. *AIP Conf. Proc.*, (CERN-LHC-Project-Report-499), Aug 2001.

[34] Verweij, A. P. and Buchsbaum, L. Experimental results of current distribution in Rutherford-type LHC cables. *Cryogenics*, 40(8-10):663–670, 2000.

[35] Lebrun, Ph. Cryogenics for the Large Hadron Collider. Technical Report CERN-LHC-Project-Report-338, CERN, Geneva, Dec 1999.

[36] Wyss, C. LHC Arc Dipole Status Report. *Particle Accelerator Conference Proc.*, pages 149–153, 1999.

[37] Fessia, P. and Perini, D. and Russenschuck, S. and Völlinger, C. and Vuillermet, R. and Wyss, C. Selection of the Cross-Section Design for the LHC Main Dipole. *IEEE Trans. Appl. Supercond.*, 10(CERN-LHC-Project-Report-347), Dec 1999

[38] Wyss, C. The LHC Magnet Programme: From Accelerator Physics Requirements to Production in Industry. *EPAC 2000 Conference Proc.*, pages 207–211, 2000.

[39] Modena, M. and Bajko, M. and Bottura, L. and Buzic, M. and Fessia, P. and Pagano, O. and Perini, D. and Pugnat, P. and Sanfilippo, S. and Savary, F. and Scandale, W. and Siemko, A. and Spigo, G. and Todesco, E. and Vanenkov, I. and Vlogaert, J. and Wyss, C. Final Prototypes, First Pre-series Units and Steps Towards Series Production of the LHC Main Dipoles. (CERN-LHC-Project-Report-487), Aug 2001.

[40] Gröbner, O. Overview of the LHC vacuum system. *Vacuum*, 60(1-2):25–34, 2001.

[41] Angerth, B. and Bertinelli, F. and Brunet, J.-C. and Calder, R. and Caspers, F. and Cruikshank, P. and Dalin, J.-M. and Gröbner, O. and Kos, N. and Mathewson, A. and Poncet, A. and Reymermier, C. and Ruggiero, F. and Scholz, T. and Sgobba, S. and Wallén, E The Large

Hadron Collider Vacuum System. *Particle Accelerator Conference, Dallas, Texas, USA*, May 1995.

[42] Gröbner, O. Vacuum system for LHC. *Vacuum*, 46(8-10):797–801, 1995.

[43] Gröbner, O. The LHC Vacuum System. (CERN-LHC-Project-Report-181), May 1998.

[44] Gröbner, O. LHC vacuum system. *Vacuum in accelerators. Proceedings. CERN Accelerator School 1999 (CAS)*, (CERN 99-05):291–306, 1999.

[45] Benvenuti, C. and Escudeiro Santana, A. and Ruzinov, V. Ultimate pressures achieved in TiZrV sputter-coated vacuum chambers. *Vacuum*, 60(1-2):279–284, 2001.

[46] Wilson, E. J. N. *An Introduction to Particle Accelerators*. Oxford Univ. Press, Oxford, 2001.

[47] Caspers, F. and Pugnat, P. and Rathjen, C. and Russenschuck, S. and Siemko, A. Currents in, Forces on and Deformations/Displacements of the LHC Beam Screen Expected during a Magnet Quench. (CERN-LHC-Project-Report-489), Aug 2001.

[48] Cruikshank, P. and Artoos, K. and Bertinelli, F. and Brunet, J. C. and Calder, R. and Campedel, C. and Collins, I. R. and Dalin, J. M. and Feral, B. and Gröbner, O. and Kos, N. and Mathewson, A. G. and Nikitina, L. I. and Nikitin, I. N. and Poncet, A. and Reymermier, C. and Schneider, G. and Sexton, J. C. and Sgobba, S. and Valbuena, R. and Veness, R. J. M. Mechanical Design Aspects of The LHC Beam screen. (CERN-LHC-Project-Report-128), Jul 1997.

[49] Artoos, K. and Cruikshank, P. and Kos, N. Mechanical and thermal measurements on a 11 m long beam screen in the LHC Magnet Test String during RUN 3A. Technical Report LHC-PROJECT-NOTE-178, CERN, Geneva, Feb 1999.

[50] Caspers, F. and Morvillo, M. and Ruggiero, F. and Tan, J. Surface Resistance Measurements and Estimate of the Beam-Induced Resistive Wall Heating of the LHC Dipole Beam Screen. Technical Report CERN-LHC-Project-Report-307, CERN, Geneva, Aug 1999.

[51] Caspers, F. and Mostacci, A. and Palumbo, L. and Ruggiero, F. Image Currents in Azimuthally Inhomogeneous Metallic Beam Pipes. (CERN-LHC-Project-Report-493), Aug 2001.

[52] Gröbner, O. and Mathewson, A. G. and Störi, H. and Strubin, P. and Souchet, R. Studies of photon induced gas desorption using synchrotron radiation. *Vacuum*, 33(7):397–406, 1983.

[53] Williams, E. M. and Le Normand, F. and Hilleret, N. and Dominichini, G. Studies of photon induced desorption of surface gas within an aluminium vacuum chamber using an X-ray source. *Vacuum*, 35(3):141–148, 1985.

[54] Gröbner, O. and Mathewson, A. G. and Marin, P. C. Gas desorption from an oxygen free high conductivity copper vacuum chamber by synchrotron radiation photons. *Journal of Vacuum Science & Technology A: Vacuum, Surfaces, and Films*, 12(3):846–853, 1994.

[55] Anashin, V. V. and Malyshev, O. B. and Osipov, V. N. and Maslennikov, I. L. and Turner, W. C. Investigation of synchrotron radiation-induced photodesorption in cryosorbing quasi-closed geometry. *Journal of Vacuum Science & Technology A: Vacuum, Surfaces, and Films*, 12(5):2917–2921, 1994.

[56] Anashin, V. V. and Derevyankin, G. and Dudnikov, V. G. and Malyshev, O. B. and Osipov, V. N. and Foerster, C. L. and Jacobsen, F. M. and Ruckman, M. W. and Strongin, M. and Kersevan, R. and Maslennikov, I. L. and Turner, W. C. and Lanford, W. A. Cold beam tube photodesorption and related experiments for the Superconducting Super Collider Laboratory 20 TeV proton collider. *Journal of Vacuum Science & Technology A: Vacuum, Surfaces, and Films*, 12(4):1663–1672, 1994.

[57] Baglin, V. and Collins, I. R. and Gröbner, O. and Grünhagel, C. and Jenninger, B. Molecular desorption by synchrotron radiation and sticking coefficient temperatures for H_2, CH_4, CO and CO_2. *Vacuum*, 67:421–428, 2001.

[58] Calder, R. and Gröbner, O. and Mathewson, A. G. and Anashin, V. V. and Dranichnikov, A. and Malyshev, O. B. Synchrotron radiation induced gas desorption from a Prototype Large Hadron Collider beam screen at cryogenic temperatures. *Journal of Vacuum Science & Technology A: Vacuum, Surfaces, and Films*, 14(4):2618–2623, 1996.

[59] Anashin, V. V. and Malyshev, O. B. and Calder, R. and Gröbner, O. A study of the photodesorption process for cryosorbed layers of H_2, CH_4, CO or CO_2 at various temperatures between 3 and 68 K. *Vacuum*, 53(1-2):269–272, 1999.

[60] Anashin, V. V. and Malyshev, O. B. and Collins, I. R. and Gröbner, O. Photon-stimulated desorption and the effect of cracking of condensed molecules in a cryogenic vacuum system. *Vacuum*, 60(1-2):15–24, 2001.

[61] Collins, I. R. and Gröbner, O. and Malyshev, O. B. and Rossi, A. and Strubin, P. and Veness, R. J. M. Vacuum Stability for Ion Induced Gas Desorption. Technical Report CERN-LHC-Project-Report-312, CERN, Geneva, Oct 1999.

[62] Malyshev, O. B. The ion impact energy on the LHC vacuum chamber walls. *EPAC 2000 Conference Proc.*, pages 951–953, 2000.

[63] Malyshev, O. B. The energy of the ions bombarding the vacuum chamber walls. Technical Report Vacuum Technical Note 99-17, CERN, Geneva, Nov 1999.

[64] Achard, M. H. Désorption des gaz induite par des électrons et des ions de l'acier inoxydable, du cuivre OFHC, du titane et de l'aluminium purs. Technical Report CERN-ISR-VA-76-34, CERN, Geneva, Aug 1976.

[65] Achard, M. H. and Calder, L. and Mathewson, A. G. The temperature dependence of the electron and ion induced gas desorption coefficients of some technological materials. Technical Report CERN-ISR-VA-78-2, CERN, Geneva, 1978.

[66] Turner, W. C. Ion desorption stability in superconducting high energy physics proton colliders. *Journal of Vacuum Science & Technology A: Vacuum, Surfaces, and Films*, 14(4):2026–2038, 1996.

[67] Barnard, J. C. and Bojko, I. and Hilleret, N. Desorption of H_2 and CO_2 from Cu by H_2^+ and Ar^+ Ion Bombardment. Technical Report LHC-Project-Note-44, CERN, Geneva, 1996.

[68] Lyneis, C. and Kneisel, P. and Stoltz, O. and Halbritter, J. On the role of electrons in RF breakdown. *Magnetics, IEEE Transactions on*, 11(2):417–419, Mar 1975.

[69] William, P. Vacuum breakdown and surface coating of RF cavities. *Journal of Applied Physics*, 56(5):1546–1547, 1984.

[70] Pivi, M. *Beam Induced Electron Multipacting in the CERN Large Hadron Collider Accelerator LHC*. PhD thesis, Università degli studi di Torino, Torino, 2000.

[71] Collins, I. R. and Gröbner, O. and Hilleret, N. and Jimenez, J. M. and Pivi, M. Electron cloud potential remedies for the vacuum system of the SPS. *10th Workshop on LEP-SPS Performance, Chamonix, France*, pages 150–154, 2000.

[72] Arduini, G. and Cornelius, K. and Gröbner, O. and Hilleret, N. and Höfle, W. and Jimenez, J. M. and Laurent, J. M. and Moulard, G. and Pivi, M. and Weiss, K. Electron cloud: Observations with LHC-type beams in the SPS. *EPAC 2000 Conference Proc.*, pages 939–941, 2000.

[73] Arduini, G. and Baglin, V. and Brünig, O. and Cappi, R. and Caspers, F. and Collier, P. and Collins, I. R. and Cornelius, K. and Garoby, R. and Gröbner, O. and Henrist, B. and Hilleret, N. and Höfle, W. and Jimenez, J. M. and Laurent, J. M. and Linnecar, T. and Mercier, E. and Pivi, M. and Ruggiero, F. and Rumolo, G. and Scheuerlein, C. and Tuckmantel, J. and Vos, L. and Zimmermann, F. Elecytron cloud effects in the CERN SPS and LHC. *EPAC 2000 Conference Proc.*, pages 259–261, 2000.

[74] Brüning, O. S. Simulations for the Beam-induced Electron Cloud in the LHC Beam Screen with Magnetic Field and Image Charges. Technical Report CERN-LHC-Project-Report-158, CERN, Geneva, Nov 1997.

[75] Furman, M. A. The Electron-Cloud Effect in the Arcs of the LHC. Technical Report CERN-LHC-Project-Report-180, CERN, Geneva, May 1998.

[76] Stupakov, G. Photoelectrons and Multipacting in the LHC: Electron Cloud Build-up. Technical Report CERN-LHC-Project-Report-141, CERN, Geneva, Oct 1997.

BIBLIOGRAPHY

[77] Baglin, V. and Collins, I. and Henrist, B. and Hilleret, N. and Vorlaufer, G. A Summary of Main Experimental Results Concerning the Secondary Electron Emission of Copper. Technical Report CERN-LHC-Project-Report-472, CERN, Geneva, Aug 2001.

[78] Baglin, V. and Bojko, I. and Gröbner, O. and Henrist, B. and Hilleret, N. and Scheuerlein, C. and Taborelli, M. The secondary electron yield of technical materials and its variation with surface treatments. *EPAC 2000 Conference Proc.*, pages 217–221, 2000.

[79] Bojko, I. and Dorier, J.-L. and Hilleret, N. and Scheuerlein, Ch. Lowering the secondary electron yield of technical copper surfaces by strong oxydation. Technical Report LHC Vacuum Technical Note 97-19, CERN, Geneva, June 1997.

[80] Bojko, I. and Hilleret, N. and Scheuerlein, Ch. Influence of air exposures and thermal treatments on the secondary electron yield of copper. *Journal of Vacuum Science & Technology A: Vacuum, Surfaces, and Films*, 18(3):972–979, 2000.

[81] Hilleret, N. and Baglin, V. and Collins, I. and Gröbner, O. and Henrist, B. and Vorlaufer, G. The secondary electron yield of copper: New experimental results and their implications. *International Workshop on Two-Stream Instabilities in Particle Accelerators and Storage Rings, KEK, Tsukuba, Japan*, 2001.

[82] Ding, M.Q. and Williams, E.M. Electron stimulated desorption of gases at technological surfaces of aluminium. *Vacuum*, 39(5):463–469, 1989.

[83] J. Gómez-Goñi, J. and Mathewson, A. G. Temperature dependence of the electron induced gas desorption yields on stainless steel, copper, and aluminum. *Journal of Vacuum Science & Technology A: Vacuum, Surfaces, and Films*, 15(6):3093–3103, 1997.

[84] Billard, F. and Hilleret, N. and Vorlaufer, G. Some results on the electron induced desorption yield of OFHC copper. Technical Report LHC Vacuum Technical Note 00-32, CERN, Geneva, Dec 2000.

[85] Rossi, A. VASCO (VAcuum Stability COde): multi-gas code to calculate gas density profile in a UHV system. Technical Report CERN-LHC-Project-Note-341, CERN, Geneva, Mar 2004.

[86] Hilleret, Noël. Non-Thermal Outgassing. *Vacuum in accelerators. Proceedings. CERN Accelerator School 2006 (CAS)*, CERN-2007-003:87–116, 2007.

[87] Yamamura, Y. and Kimura, H. Particle reflection and ion-induced desorption from tungsten surfaces with chemisorbed nitrogen. *Surface Science Letters*, 185(1-2):L475–L482, 1987.

[88] Taglauer, E. and Beitat, U. and Marin, G. and Heiland, W. Sputtering of adsorbed layers by ion bombardment. *Journal of Nuclear Materials*, 63:193–198, 1976.

[89] Robinson, M. T. and Torrens, I. M. Computer simulation of atomic-displacement cascades in solids in the binary-collision approximation. *Phys. Rev. B*, 9(12):5008–5024, 1974.

[90] Biersack, J.P. and Haggmark, L.G. A Monte Carlo computer program for the transport of energetic ions in amorphous targets. *Nuclear Instruments and Methods*, 174(1-2):257–269, 1980.

[91] Johnson, R. E. and Schou, J. Sputtering of Inorganic Insulators. *Mat.-Fys. Medd. Danske Vid. Selsk.*, 43:403–494, 1993.

[92] Chen, YiPing. Monte Carlo simulation of gas desorption process induced by low-energy ions. *Journal of Nuclear Materials*, 303(2-3):99–104, 2002.

[93] Schou, J. Slowing-down processes, energy deposition, sputtering and desorption in ion and electron interactions with solids. *Vacuum in accelerators. Proceedings. CERN Accelerator School 2006 (CAS)*, CERN-2007-003:169–178, 2007.

[94] Lindhard, J. and Nielsen, V. and Scharff, M. and Thomsen, P. V. Integral equations governing radiation effects. *Mat.-Fys. Medd. Danske Vid. Selsk.*, 33(10), 1963.

[95] Bohr, N. The penetration of atomic particles through matter. *Mat.-Fys. Medd. Danske Vid. Selsk.*, 18(8), 1948.

[96] Niehus, H. and Heiland, W. and Taglauer, E. Low-energy ion scattering at surfaces. *Surface Science Reports*, 17(4-5):213–303, 1993.

[97] Gnaser, H. *Low-energy ion irradiation of solid surfaces*. Springer Tracts in Modern Physics. Springer, Berlin, 1999.

[98] Wilson, W. D. and Haggmark, L. G. and Biersack, J. P. Calculations of nuclear stopping, ranges, and straggling in the low-energy region. *Phys. Rev. B*, 15(5):2458–2468, 1977.

[99] Lindhard, J. and Nielsen, V. and Scharff, M. Approximation method in classical scattering by screened coulomb fields. *Mat.-Fys. Medd. Danske Vid. Selsk.*, 36(10), 1968.

[100] Seah, M. P. and Clifford, C. A. and Green, F. M. and Gilmore, I. S. An accurate semi-empirical equation for sputtering yields I: for argon ions. *Surface and Interface Analysis*, 37(5):444–458, 2005.

[101] Seah, M. P. An accurate semi-empirical equation for sputtering yields, II: for neon, argon and xenon ions. *Nuclear Instruments and Methods in Physics Research Section B: Beam Interactions with Materials and Atoms*, 229(3-4):348–358, 2005.

[102] Lindhard, J. and Scharff, M. Energy Dissipation by Ions in the keV Region. *Phys. Rev.*, 124(1):128–130, 1961.

[103] Sigmund, P. *Particle penetration and radiation effects: general aspects and stopping of swift point charges.* Springer Series in Solid State Sciences. Springer, Berlin, 2006.

[104] NPL–UK's National Measurement Laboratory. Sputter Yield Values. http://www.npl.co.uk/server.php?show=ConWebDoc.605, August 2008.

[105] Dahl, D. A. *SIMION 3D Version 7.0 User's Manual.* Bechtel BWXT, Idaho, 2000.

[106] Blaum, K. *Laserspektroskopie, Fallen und ihre Anwendungen - lecture notes.* Johannes Gutenberg - Universität Mainz, Mainz, 2006.

[107] Freund, R. S. and Wetzel, R. C. and Shul, R. J. and Hayes, T. R. Cross-section measurements for electron-impact ionization of atoms. *Phys. Rev. A*, 41(7):3575–3595, 1990.

[108] Benvenuti, C. and Hauer, M. Low pressure limit of the bayard-alpert gauge. *Nuclear Instruments and Methods*, 140(3):453–460, 1977.

[109] Paschen, F. Über die zum Funkenübergang in Luft, Wasserstoff und Kohlensäure bei verschiedenen Drucken erforderliche Potentialdifferenz. *Annalen der Physik*, 273(5):69–75, 1889.

[110] Raizer, Y. P. *Gas discharge physics. Fizika gazovogo razryada.* Springer, Berlin, 1991. Translated from Russian by Vitaly I Kisin.

[111] Hablanian, M. H. *High-Vacuum Technology: A Practical Guide; 2nd Ed.* Mechanical Engineering Series. Taylor and Francis, Boca Raton, FL, 1997.

[112] Sagara, A. and Kamada, K. Compilation and evaluation of ion impact desorption cross-section. *Journal of Nuclear Materials*, 111-112:812–815, 1982.

[113] Ziegler, J. SRIM – The Stopping and Range of Ions in Matter. http://www.srim.org/, August 2008.

[114] Andersen, H. H. and Bay, H. L. Heavy-ion sputtering yields of gold: Further evidence of nonlinear effects. *Journal of Applied Physics*, 46(6):2416–2422, 1975.

[115] Chernyaev, A. V. Modeling Molecular-Ion Implantation into Solids. *Russian Microelectronics*, 32(1):21–25, 2002.

[116] Burnett, J. W. and Pellin, M. J. and Calaway, W. F. and Gruen, D. M. and Yates, J. T. Ion dose dependence of the sputtering yield of Ru(0001) at very low fluences. *Phys. Rev. Lett.*, 63(5):562–565, 1989.

[117] El-Said, A. S. and Heller, R. and Meissl, W. and Ritter, R. and Facsko, S. and Lemell, C. and Solleder, B. and Gebeshuber, I. C. and Betz, G. and Toulemonde, M. and Möller, W. and Burgdörfer, J. and Aumayr, F. Creation of Nanohillocks on CaF_2 Surfaces by Single Slow Highly Charged Ions. *Physical Review Letters*, 100(23):237601, 2008.

[118] J. W. Burnett, J. W. and Biersack, J. P. and Gruen, D. M. and Jørgensen, B. and Krauss, A. R. and Pellin, M. J. and Schweitzer, E. L. and Yates Jr., J. T. and Young, C. E. Depth of origin of sputtered atoms: Experimental and theoretical study of Cu/Ru(0001). *Journal of Vacuum Science & Technology A: Vacuum, Surfaces, and Films*, 6(3):2064–2068, 1988.

[119] Panzner, G. and Egert, B. and Schmidt, H.P. The stability of CuO and Cu_2O surfaces during argon sputtering studied by XPS and AES. *Surface Science*, 151(2-3):400–408, 1985.

[120] Müller, A. and Benninghoven, A. Investigation of surface reactions by the static method of secondary ion mass spectrometry: V. The oxidation of titanium, nickel, and copper in the monolayer range. *Surface Science*, 41(2):493–503, 1974.

[121] Benninghoven, A. and Müller, A. Investigation of surface reactions by the static method of secondary ion mass spectrometry: II. The oxidation of chromium in the monolayer range. *Surface Science*, 39(2):416–426, 1973.

[122] Mathewson, A. G. ISR cleaning procedure tests. Technical Report CERN-ISR-VA-74-10, CERN, Geneva, Feb 1974.

[123] Lozano Bernal, María Pilar. *Estudio de la desorción estimulada por iones de adsorbatos en superficies de interés tecnológico (Al, Cu)*. PhD thesis, Madrid Univ., Madrid, 2004. Presented on 10 May 2004.

[124] Tratnik, H. and Störi, H. and Hilleret, Noël. *Electron Stimulated Desorption of Condensed Gases on Cryogenic Surfaces*. PhD thesis, Wien TU, Wien, 2005. Presented on 01 Sep 2005.

Die VDM Verlagsservicegesellschaft sucht für wissenschaftliche Verlage abgeschlossene und herausragende

Dissertationen, Habilitationen, Diplomarbeiten, Master Theses, Magisterarbeiten usw.

für die kostenlose Publikation als Fachbuch.

Sie verfügen über eine Arbeit, die hohen inhaltlichen und formalen Ansprüchen genügt, und haben Interesse an einer honorarvergüteten Publikation?

Dann senden Sie bitte erste Informationen über sich und Ihre Arbeit per Email an *info@vdm-vsg.de*.

Sie erhalten kurzfristig unser Feedback!

VDM Verlagsservicegesellschaft mbH
Dudweiler Landstr. 99
D - 66123 Saarbrücken

Telefon +49 681 3720 174
Fax +49 681 3720 1749

www.vdm-vsg.de

Die VDM Verlagsservicegesellschaft mbH vertritt

Printed by Books on Demand GmbH, Norderstedt / Germany